Contributions to Economics

For further volumes:
http://www.springer.com/series/1262

Małgorzata Runiewicz-Wardyn

Knowledge Flows, Technological Change and Regional Growth in the European Union

Springer

Małgorzata Runiewicz-Wardyn
Kozminski University
Warsaw
Poland

ISSN 1431-1933
ISBN 978-3-319-03335-8 ISBN 978-3-319-00342-9 (eBook)
DOI 10.1007/978-3-319-00342-9
Springer Cham Heidelberg New York Dordrecht London

Printed on acid-free paper

Springer is part of Springer Science+Business Media (www.springer.com)

Acknowledgments

My most profound thanks to Prof. Grzegorz W. Kolodko, Director of the Transformation, Integration and Globalization Economic Research Center (TIGER), who gave me the opportunity to pursue my academic career and scientific interests. I would also like to express my gratitude to numerous experts and colleagues from Harvard University, the University of California at Berkeley, Suffolk University, and Université libre de Bruxelles for sharing their precious comments and experience, and giving their professional advice on my research. For all the interesting comments and valuable feedback my special thanks go to Prof. Philippe Aghion and Prof. Edward L. Glaeser of Harvard University; Prof. David Dowall, Prof. Stephen S. Cohen, and Prof. John Zysman of the University of California at Berkeley; Prof. Zbigniew Bochniarz, the Evans School of Public Affairs, University of Washington; and Prof. Robert DeFillippi of Suffolk University in Boston. I am also very grateful to Prof. Henry Capron and Prof. Michele Cincera from Université libre de Bruxelles for hosting me for two short-term exchange visits during which I had the opportunity to have long discussions with people at the root of my interest. It also enabled me to conduct interviews with representatives of the Wallonie and Flanders regions' technological clusters as well as policy makers in European institutions. Many thanks also to Dr. Jaan Masso, Dr. Alessandro Vecciato, and Dr. Lydia Greunz for their technical support and feedback.

I would also like to thank my interviewees from different scientific fields for their expert feedback and comments. Finally, I would like to express my deepest and most heartfelt gratitude to my husband, my son, and my parents for their patience and encouragement. Thank you.

Preface

Knowledge and learning have always been central driving forces of economic growth and prosperity. Different speeds of knowledge accumulation and diffusion, as well as different socio-political environments have hampered the successful absorption of knowledge and transfer into economic resources. Almost a century ago, Alfred Marshall embraced this interlinkage in one of his statements (1920, pp. 270–271): "Nearly all important knowledge has long deep roots stretching downwards to distant times; and so widely spread have been these roots, so ready to send up shoots of vigorous life, that there is perhaps no part of the old world in which there might not long ago have flourished many beautiful and highly skilled industries, if their growth had been favored by the character of the people; and by their social and political institutions (. . .)." In fact, people are inherently creative and innovative and that if only the institutional and political circumstances were right, their knowledge would be used for continued technological progress. The economic utility of knowledge rests on its creation, diffusion, and technological application. Knowledge and technology are like the "heart" and "mind" of every economy. By integrating knowledge factors with technological advancement and inserting them into the aggregate production function, economies can take full advantage of scale and long-term prosperity.

Today, more than ever before in human history, there is greater appreciation for the importance of knowledge and learning. Accelerating technological advancement and globalization puts continuous pressure on learning and constantly upgrading skills. Therefore, it should be every government's role to provide necessary technological infrastructure, encourage and support business R&D efforts, and promote lifelong learning. Increasing the quality of human resources improves the efficiency of the workforce, raises aggregate productivity, and improves allocative efficiencies of economic resources that could generate growth for future generations.

Contents

List of Figures

List of Maps

List of Tables

List of Annexes

Annex 1

Annex 2

Introduction

The "knowledge economy," induced by technological advances, has been a powerful engine for economic growth and competitiveness of world economies. Knowledge and the innovative capacity of human beings have become factors of production. Traditional production factors, such as capital and labor, are no longer sufficient to enable successful competition in the context of rapid technological progress and globalization. Innovation and technological advances in economically important areas, such as electronics, computers, telecommunications, and biotechnologies, and particularly in information and communication technologies (ICT), have brought and will continue to bring pervasive impacts on local, regional, and national economic systems.

However, the precise investigation of connections between knowledge, technological advances, and economic growth has been hampered by a lack of robust measures of technological change and its relationship to knowledge spillovers and economic performance (productivity and GDP growth). Paul Romer (1990, 1993) was one of the first to test the relationship between endogenous technological change, knowledge spillovers, and economic growth. Romer, along with the group of "new (endogenous) growth" theorists, showed that economic growth depends on both access to new technological ideas and diffusion of these ideas through the productive structure (spillover externalities). Despite the growing empirical literature on successful cases of the knowledge (innovation)-driven growth of regions, the rate of introduction, adoption, and diffusion of new technologies across different parts of the world is not even, and so the capacity to turn knowledge and innovation into regional growth is different among regions. This has made theorists look beyond the technological dimension of growth and introduce three other important dimensions – institutional, time, and space. In fact, the creation of innovative products and services depends not only on the productivity of research activities of scientists and engineers, but also on the communities and their institutions as well as the time frame in which they live and work. Many research findings have provided evidence of the importance of geographical proximity, and of the development of geographically concentrated clusters of institutions in new knowledge creation, which has given the directions of development of the "new economic geography." Regions, rather than states, with their knowledge base, their innovative firms and institutional interlinkages, are the key contributors to innovations (Keeble and Wilkinson, 2000; Bresnaham and Gambardella, 2004). The founding of new firms occurs in a limited number of regions with access to leading research institutions, venture capital, and an abundant pool of educated labor (Owen-Smith et al., 2002). Almost one third of all high-tech patents in the EU are concentrated in four regions: Oberbayern (Germany), Île de France (France), Noord-Brabant (the Netherlands), and Etelä-Suomi (Finland). Almost half of EU biotechnology patenting is located in Île de France (France), Oberbayern (Germany), and Denmark (Eurostat, 2008). This trend repeats itself on a global scale. For example, slightly more than half of the biotechnology firms in the United States are clustered in just three areas – Cambridge (Massachusetts), San Diego County, and the San Francisco Bay Area in California.

These regions were much more successful in providing a "critical mass" for encouraging research and enhancing networking between business, higher education institutions, research centers, and technology parks. As a result, they follow the virtuous circle of growth – innovations and technological change are fueling economic growth, which is being fed back into higher earnings and greater investment in education and R&D.

The following book attempts to contribute to the existing empirical findings and theoretical discussion on the interlinkages between determined technological change, knowledge spillovers, and economic growth of regions. This would not be possible at the state or country level, since their size is generally too large to capture geographically and technologically mediated knowledge spillovers.

Based on the existing selected literature, the following book addresses two sets of questions:

Empirical (A): What is the heterogeneity of the EU's geographical landscape in terms of innovative potential, technological advancement, and regional growth? What is the role of institutional, technological, and geographical dimensions in explaining the patterns of regional innovative performances and knowledge spillovers in the EU? Which agglomeration externalities explain technological dynamics in the EU's high-tech industries? What is the role of knowledge spillovers and technological change in regional growth and catching up in the EU?

Theoretical (B): How well do the systemic and evolutionary approaches to regional innovation processes explain the present innovative performance of regions? What is the role of geographic and technological proximities in explaining research and patenting activity of regions? What types of dynamic externalities define technological advancement in high-tech industries? How well do endogenous growth and technological gap theories explain key empirical regularities in the growth of regions?

The first question in the theoretical section aims to advance a theoretical framework of regional innovation systems (RIS) toward an evolutionary approach to innovation processes. The concept of RIS has gained much interest among academic researchers and policy makers in past decades. It has presented a new and promising analytical framework for advancing our understanding of the innovation process in the regional economy (Asheim et al., 2003; Cooke et al., 2002). However, critics of RIS claim it gives only a partial view of formal research and development (R&D) activities and does not consider a dynamic perspective of innovation processes, which will evaluate the innovative behavior of regions from the perspective of their institutional setup and by historical factors (Mytelka and Farinelli, 2000; Seifoddin et al., 2008; Braczyk et al., 1998). This issue is especially significant for the assessment of the EU's regional innovative performance. The main hypothesis is that the differences in innovation performance between EU-15[1] and EU-10[2] Member States regions result from the differences in their socio-political environments. The knowledge flows within the RISs of EU-10 are still conditional upon public R&D support, and follow the neo-institutional model of university-industry-government interlinkages. This is in contrast to many of the EU-15 regions that have developed generative R&D relationships and a spontaneous convergence of the industrial and academic worlds.

Furthermore, this book will try to contribute to the ongoing discussion on the role of space and technological proximity in the process of knowledge spillovers and innovation. The basic hypothesis behind the second question in the theoretical section is that the innovative capacity of the EU regions is determined by both geographically and technologically mediated knowledge spillovers. Despite the growing number of empirical studies, evidence of geographical knowledge spillovers is very fragmented and devoted almost entirely to the experience of advanced regional economies, with little distinction of the type of knowledge and consideration of the sector's specific trends. The interview results presented in Chapter IV on the role of universities in local knowledge flows in high-tech fields will challenge the role of distance for knowledge spillovers later on, and thus the prime reasoning of "new economic geography" theorists, by incorporating the experiences of R&D networks of the European Research Area (ERA).

New evidence of geographically and technologically mediated knowledge spillovers, combined with sectoral analyzes of local technological externalities, may elicit further discussion on the role of place-specific characteristics in technological change. It is now commonly considered that the clustering of innovation activities, especially in high-tech industries, is an important base for knowledge externalities. Agglomeration externalities may result from specialized (suppliers and knowledge base) or diversified (shared or complementary suppliers and knowledge base) regional structures. This book revisits the central part of MAR's, Jacobs' (1969), and Porter's (competition) externalities in explaining the regional

[1] EU-15 Member States include Austria, Belgium, Luxembourg, Denmark, France, Germany, Ireland, Italy, the Netherlands, Spain, Portugal, Greece, Sweden, Finland, and the United Kingdom. They joined prior to 2004.

[2] EU-10 new Member States include – Czech Republic, Estonia, Cyprus, Latvia, Lithuania, Hungary, Malta, Poland, Slovenia, and Slovakia. They joined the EU after 2004, except for the two latest Member States: Bulgaria and Romania.

innovation process by testing the hypothesis that the presence of dynamic externalities (MAR, Jacobs', and Porter's externalities) varies with the industry's technological field, its interdisciplinary character, and its technological maturity.

The third question in the theoretical section aims to contribute on the role of technological advancement and knowledge spillovers in regional growth. Most theories, like the "milieu innovateur theory" and the "learning region" or "regional innovation systems" theory, assume a direct linkage between knowledge produced and regional economic performance. All these theories share the idea that there is a positive relation between knowledge created in a particular region and the region's economic performance. The links between technology and economic growth are increasingly better understood, but the role of technological spillovers across space in regional catching up still remains enigmatic. The book considers the relationship between technological change in a local economy and its territorial neighbors. The final hypothesis tested in this book is that regional growth in the EU is related to both exogenously and endogenously determined technological change. The rapid accumulation of capital investments (exogenously determined technological change) pushes up relative demand for skilled labor and generates the process of learning (endogenously determined technological change). The latter is especially true for the EU-10 Member States regions.

This book aims to provide satisfying answers to the above-mentioned empirical and theoretical questions by investigating the EU-25 Member States regional landscape. The EU-25 (later defined as EU) refers to all the EU current Member States regions, except for Bulgaria and Romania.[3]

The regions are analyzed on NUTS2 regional level,[4] with a few exceptions: Ireland, Denmark, and Germany, for which data was available only on the national and NUTS1 level. Very few attempts have been made to empirically investigate the process of knowledge creation and spillovers within the European regions to date, especially by including recently joined Member States regions. Most studies in this field are carried out for single European countries, EU-15 regions, or other advanced economies such as the United States or Canada.

This book proceeds as follows: there are six chapters, including this Introduction. Chapter I aims to outline some theoretical frameworks for knowledge as a driver of technological change, its measures and relationship to regional spatial and growth dynamics. The chapter begins with an explanation of the concept of knowledge and innovation and its role in technological change. Based on a critical literature review, it identifies mechanisms of knowledge creation and spillovers, the transition of knowledge to innovations and technological change, and the role of

[3] The latter two states joined the EU in 2007; therefore, the data for many economic, social, and technical indicators for these countries are not available on a regional level.

[4] The NUTS (*Nomenclature of Territorial Units for Statistics*) is based on the existing national administrative subdivisions of the EU geographical landscape for statistical purposes. The NUTS regulation lays down a minimum population size of three million and a maximum size of seven million for the average NUTS-1 region, and a minimum of 800,000 and a maximum of three million for NUTS-2 regions, http://www.europa.eu/

geographical, technological, and institutional space in innovation processes, taking into account the key actors involved in knowledge creation as well as their potential feedback relations. Chapter II provides some useful facts and figures that highlight the heterogeneity of the European regional landscape regarding wealth and knowledge-creation capacities. Different kinds of indicators are examined. Firstly, the positioning of European regions in terms of GDP per capita is realized. Secondly, we attempt to identify the respective knowledge absorption, transfer, and diffusion capacities of the EU regions by looking at patent data, R&D expenditures, and human qualification levels. In sum, Chapter II aims to provide a clear picture of the current innovative potential and technological capabilities of European regions. This should enable a better understanding of the estimation results obtained in Chapter III and their resulting policy implications. Chapter III attempts to achieve some conclusions on the relative efficiencies of regional innovation systems (RIS) of the EU Member States, including respective contributions of their university, business, and government R&D expenditures, high-tech manufacturing, and knowledge-intensive services in the process of knowledge creation. The third chapter aims to determine what the university-business R&D interlinkages are and what role human qualifications play in the process of knowledge creation. Chapter IV investigates geographically and technologically determined knowledge spillovers among European regions, and thus contributes to the understanding of the diffusion processes of innovation and knowledge in space. Chapter V discusses how knowledge externalities are being shaped by the composition and organization of regional productive systems and whether these externalities influence knowledge creation. The role of regional productive systems in generating technological externalities may differ when a particular high-tech industry and its technological life cycle are considered. Taking regional as well as sector-specific variables into account, this chapter provides some interesting insights regarding the influence of different kinds of externalities on innovations in the six high-tech industries. Chapter VI analyzes the role of technological change and knowledge spillovers in the technological catching up and economic growth processes. The book ends with a number of conclusions for policy and theory and Annexes.

Chapter 1
Knowledge as a Driver of Technological Change and Regional Growth

1.1 Introduction

In order to pursue a comprehensive analytical approach to study technological change and its role in regional growth in the EU, this chapter draws on an overview of the economic theory of knowledge, endogenous growth theories, and "new economic geography" to prepare a theoretical framework for the models estimated in Chaps. 3, 4, 5 and 6.

The role of knowledge spillovers and technological change in regional economic growth is particularly difficult to understand. This is related to the complexity of many technological advances on the one hand, and the nature of knowledge flows, on the other. Complex technological advances require the process of innovations to combine knowledge from several scientific disciplines and interaction between various actors, especially in the high-tech fields. Another difficulty in measuring the relationship between knowledge spillovers and technological change is the fact that knowledge flows cannot be narrowed to regional or national boundaries. Knowledge spills over from its source of origin and produces increasing returns and economic growth across the broader territorial space, which has made it a major topic of discussion among regional economists and geographers. Apart from the territorial organization of knowledge production, understanding the process of knowledge spillovers and its relationship to technological change depends upon complex interactions of different knowledge types and their sources.

The following chapter presents a more integrated approach to the role of knowledge in technological change by considering its technological, spatial, scientific, and institutional dimensions, and attempts to enrich the theoretical discussion over the interlinkages between knowledge flows, innovations, and technological change, on the one hand, and technological change and regional growth on the other. The first two sections of the chapter define knowledge and discuss its types and measures. Different aspects of knowledge and its role in technological change are elaborated. The second section reviews spatial dimensions of technology, focusing on the location of innovation, diffusion of technology, knowledge networking, and localized knowledge

M. Runiewicz-Wardyn, *Knowledge Flows, Technological Change and Regional Growth in the European Union*, Contributions to Economics,
DOI 10.1007/978-3-319-00342-9_1, © Springer International Publishing Switzerland 2013

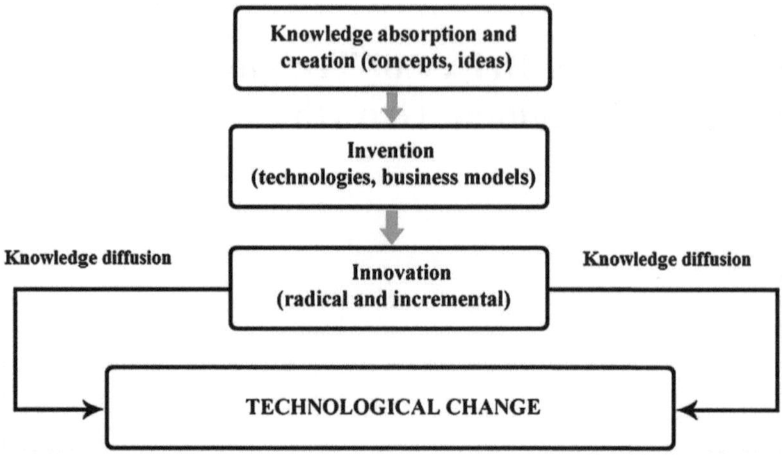

Fig. 1.1 The process of invention, innovation, and diffusion of technology (Source: Own illustration. The basis of technological advance is invention, which is the result of scientific knowledge and R&D efforts. Innovation is the successful introduction of invention in the market. The diffusion of inventions and/or innovations through imitating or copying leads to technological change in industry or society)

spillovers. Sections 1.3 and 1.4 present systemic and cluster-related approaches to technological change, its structural and spatial dimensions, and its relationship with regional growth theory. The fifth section deals with the role of technological change in economic growth by critically discussing the role of technological change in the framework of neoclassical and new (endogenous) growth theories. The chapter ends with a summary and conclusions of the most important findings.

1.2 Technological Change: Conceptual and Methodological Related Issues

1.2.1 Role of Knowledge as a Driver of Technological Change

In simple terms, technological change is a process of invention, innovation, and diffusion of technological knowledge (Fig. 1.1). Invention is the most important product of scientific knowledge; it includes the discovery of new concepts, processes, or new products that can be patented. Innovation, on the other hand, involves the use or development of an invention for some useful purpose, e.g. conversion into a viable commercial offer. Technology is the aggregation of all existing inventions and innovations. Diffusion is a process of adaptation and application of technological knowledge documented in scientific publications or patents (Crane 1972).

Technological change can refer to the continuous process of inventing new ways to improve technology and its diffusion throughout industry or society (Kolodko 2008). The accumulation of knowledge and research findings in a particular firm, sector, or locality and its application in production processes results in creating new or improving existing products and in a greater efficiency of the production processes (Marx 1973).

Rapid technological progress requires supplementing internal knowledge with knowledge acquired outside, as well as validating the internal knowledge (one's own R&D efforts of home institution) with the external one (originating from others' R&D efforts) (Capello 1999; Caloghirou et al. 2004; Capello and Faggian 2005; Belussi et al. 2002). Internal knowledge refers also to the employee skills of a particular home firm or R&D institution. The greater the stock of already accumulated knowledge (based on previous innovations) is, the greater the absorption capacity of external knowledge will be. There are important complementarities between innovative performance and absorptive capacity. Previous innovations promote absorptive capacity. This is because the generation of new ideas and the associated knowledge spillover build absorptive capacity. Previous innovations promote the absorption of cutting-edge technology and inspire new ideas and innovations (Goldberg et al. 2010). Consequently, technological change depends on the stock of already accumulated knowledge (based on previous innovations), R&D activity, and the efficiency of the knowledge absorption and diffusion processes (Grubler and Messner 1998; Goulder and Schneider 1999; van der Zwaan et al. 2002). The level of human capital (their qualifications) and the stock of already accumulated knowledge (patents, publications, etc.) could serve as a measure of absorption capacities. Whereas, according to Capron (2002) and Abramovitz (1986), the levels and rates of growth of R&D expenditures are viewed as reliable indicators of knowledge diffusion as they include direct efforts toward invention and innovation.

The innovation literature further recognizes that the role of accumulated knowledge and research findings on technological change depends on the nature of the innovation implemented: radical or incremental (Dewar and Dutton 1986; Gersick 1991; Pennings 1988; Tushman and Romanelli 1985). Radical innovation involves large technological advancements, which usually require spending large amounts of money on R&D for highly uncertain returns on investment. As a result, such innovations are capable of completely replacing old products and industries with new ones, e.g. replacing oil lamps with electric ones (Schumpeter 1942; Gersick 1991), whereas incremental innovations refine or exploit the potential of already established production processes or products. Freeman (1994) noted that the role of radical innovations in industries' performance has grown in importance throughout the last century. Industry transformation happens because the adoption of radical innovation makes old routines obsolete (Nelson and Winter 1982). For instance, Castiaux (2007) suggested that "the only possibility to guarantee a leading position is to be the first on the ball."

However, radical innovation is not an easy outcome to obtain, as it usually requires spending large amounts of money on R&D. Mowery and Rosenberg (1979) pointed out that, since the market knows little or nothing of new products, the progress of radical innovations seem to stem from the sciences and technologies originating outside of the particular industry or sector. Therefore, many of these radical innovations come from R&D labs, and the inventor of radically innovative products is not always the one who receives the benefits resulting from this innovation. They involve a substantially new technology (technological breakthrough[1]) and are especially important for the high-tech industries' emergence and survival (Chaney et al. 1991; Stephan van Dijk 2008). However, this technology- or science-push model of innovation ignores the importance of incremental innovations, which often occur in situations such as learning by doing and in user-producer relations (Fagerberg 2005, p. 9; Dodgson 2000, p. 17). In contrast, incremental innovations are important in maintaining or even improving the market shares of market leaders (Freeman 1982; Dosi 1982). For example, one of the ways to accomplish these is by introducing "refinements and extensions of established designs that result in substantial price or functional benefits to users." This is also referred to as *demand-pull model innovation* since it is stimulated by demand. Firms usually perform searches for new knowledge because they believe customer demand exists behind the products or processes (Banbury and Mitchell 1995).

Until the 1980s, the linear models of innovations dominated the innovation-related literature, with the "science-push" or the "demand-pull" models as the two dominant approaches. Yet, empirical evidence shows that innovation is a complex process involving several factors and uncertainties that do not necessarily occur in a linear way (Kline and Rosenberg 1986). It is a process that demands close coordination of adequate technical and managerial knowledge as well as excellent market judgment in order to satisfy economic, technological, and other constraints – all simultaneously. As a result, the past decades have witnessed the development of more interactive models of innovation processes representing the more complex reality of technological advancement (Manley 2002, p. 3; Morgan 1997, p. 493). The interactive model suggests that innovation does not always occur chronologically where research and science come first. Rather, innovation is a result of an interactive learning process between individuals or groups, which helps actors to recombine existing knowledge in new ways (Lundvall 1992). It also includes the organizational and experience-related aspects of technology (learning by doing) (Storper and Walker 1989, p. 50). Improvements in the organization of work often accompany the successful application of new technology (e.g. machines). The efficiency of production and the usage of new technologies may greatly depend on learning through experience.

Moreover, the interactive model of innovation implies that technological progress is best made by adapting teaching and learning processes to the capabilities of new technologies. Lifelong learning must be intimately involved in promoting technological advancement in at least two ways: firstly, by understanding the cultural

[1] "Technological breakthrough" refers to the use of a substantially different technology than existing products without considerably increasing the benefits to consumers, whereas "radical innovation" is a product that is high on both the technology and market dimension; it involves a substantially different technology while offering a substantial increase in customer benefits. See also: http://www.carlsonschool.umn.edu/Assets/71621.pdf

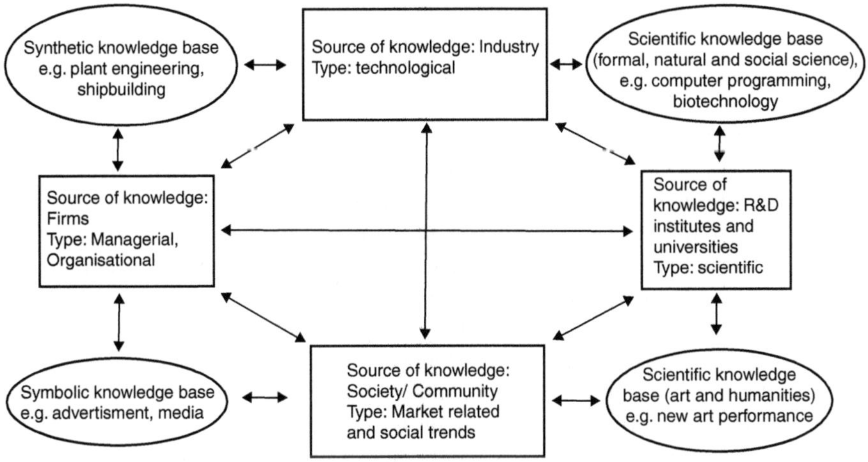

Fig. 1.2 The interactive model of the innovation process (Source: Own elaboration)

implications of technological development and, secondly, by utilising existing knowledge to develop new ways in which we could advance the latest technology. Providing learning opportunities encourages the development of innovative skills. Organizations such as "centers of excellence" provide technical assistance and training on cutting-edge innovations in their respective fields. By bringing together interdisciplinary teams of academicians, researchers, and industry representatives, these centers address important challenges of diverse high-technological areas (Abernathy and Utterback 1978; Gatignon et al. 2002).

1.2.2 Types and Sources of Knowledge in the Interactive Model of Innovations

The rapidly growing and increasingly complex research and innovation processes require a large variety of knowledge types and sources. Nonaka and Takeuchi (1995), as well as Lundvall and Borrás (1998), pointed out that the knowledge creation and exploitation processes require a dynamic interplay of various types of knowledge and the transformation of tacit and codified forms of knowledge as well as a strong interaction between people within organizations and among them. Figure 1.2 highlights knowledge creation and shows that the process of innovation is the result of interaction between various actors and types of knowledge involved in this process – scientific, technological, managerial, and market related. It demonstrates that innovation results from a combination of push and pull models, integrating the internal and external knowledge coming from different institutional spheres, including government, industry, and public sectors, and recognizing the feedback loops between them.

Although all types of knowledge are interrelated and mutually reinforcing, there are obvious benefits of using certain types of knowledge by different actors in different fields. Technological knowledge may be more important for scientific fields that use applied research (on one or several disciplines). Indeed, scientific knowledge resulting from universities and R&D institutions is an important provider of technological externalities; however, its influence on innovations and their direct application in production processes may depend on the specific type of academic field. Advances in the hard sciences (physics, chemistry, biology, mathematics, and the broad spectrum of engineering fields) and the social sciences are likely to exert a stronger effect on economic growth than advances in the medical and human sciences (Antonelli and Fassio 2012). Their contribution in the form of technological innovations is well documented. The cumulability between scientific knowledge and technological knowledge is very high as R&D activities concern mainly hard sciences (Antonelli and Fassio 2012; Stephan 2011). With regard to the social sciences, they play a central role in the introduction of organizational innovations. More efficient and effective methods of management of business and related activities (e.g. marketing, advertisement, industrial relations, and finance) rely upon advances in academic knowledge in social sciences. The ever-growing number of business schools certifies the importance of the social sciences in providing organizational knowledge and improving the efficiency of production processes in all industries and sectors of economic activity (Evangelista and Vezzani 2010). Improvements in managerial skills among engineers, scientists, and other technologists help to develop their interpersonal skills and the ability to combine and integrate different elements of R&D process and the application of new technologies in the production process. As for the medical and natural sciences, they draw on analytical science but their direct effects on innovations and their economic efficiency are limited to the medical, biotechnology, and pharmaceutical industries (Grebel 2011). Finally, the advances in art and humanities (literature, history, sociology, politics, education, psychology), even though they do not provide any direct input to technological innovations, may increase cultural tolerance and diversity of ideas that influence creativity and the innovation process.

Technological knowledge is the cornerstone of innovation (Leonard-Barton 1992) and the main source of competitive advantage in high-tech firms (Bettis and Hitt 1995; Tushman and Anderson 1986). It refers to the knowledge of manufacturing, engineering, or production methods or tools needed to serve the market.[2] Technological knowledge incorporates the level of education of the people involved, acquired from proficiency in R&D and engineering and scientific and related activities (Bierly and Chakrabarti 1996). This type of knowledge is less likely to be copied or found in competing firms because it is harder to transfer and develops based on specific investments (Nerkar and Roberts 2004). Apart from the skills learned through formal technological education, there are also skills learnt on the job through the process of "learning by doing." The efficiency of production and many

[2] Technological knowledge should not be confused with technical knowledge. The term "technology" means the practice and results of engineering as well as scientific research on engineering, whereas "technical" applies to engineering practice (Ropohl 1997, pp. 65–72).

organizational innovations may greatly depend on learning through the experience of companies' workers (Storper and Walker 1989, p. 50).

Technological knowledge can be proxied by R&D intensity, knowledge-transfer activities, the number of patents, the number of scientists working at firms, or the number of scientific publications (e.g. Thornhill 2006). University-business linkages are very important for knowledge transfer activities but they are not limited to technological experience. In fact, there are multiple knowledge-exchange mechanisms; the most important of these involves people, but also market, managerial, and organizational knowledge transfer. Understanding the market and its needs reveals new technological opportunities and enhances the ability to innovate (McKelvie 2007). The more one knows about one's customers, the better it is for innovation (Drucker 1985; Shane 2000; Gaglio and Katz 2001; Cooper et al. 1995). The study conducted by von Hippel (1986) shows that an accurate understanding of market issues is nearly essential to the successful release of new products.

Bearing in mind that knowledge is context specific, every industry will be strongly shaped by its specific knowledge base: analytical, synthetic, and symbolic (Asheim and Coenen 2005). In the case of many high-tech industries, in which radical innovations take place, knowledge creation is often based on scientific and technological (engineering) knowledge bases.[3] Both basic and applied research, as well as the systematic development of products and processes, are relevant activities for high-tech fields. Companies typically have their own R&D departments but they also rely on the research results of universities and other research organizations in their innovation processes. Thus, university-industry links and respective networks are important and more frequent than in other types of industries. Knowledge inputs and outputs are more often codified in this type of knowledge base than in other types, e.g. patents and licensing activities. However, tacit knowledge acquired through social practice is also needed in the process of new knowledge creation (Nonaka et al. 2000; Johnson and Lundvall 2001).

A synthetic knowledge base refers to industrial settings, where innovation takes place mainly through the application of existing knowledge or through new combinations of knowledge. In general, R&D is less important here than in the scientific knowledge base; it takes the form of applied research, but more often it is in the form of product or process development. University-industry links are relevant, but they are clearly more in the field of applied research (Asheim et al. 2005).

This type of knowledge base is often necessary to solve specific problems arising in interaction with clients and suppliers (Fig. 1.2). Both technological and market-related knowledge are important for the innovation product and processes' activities in these industries. However, the intra- and inter-industry links and respective networks, thus, are important and more frequent than in other types of

[3] It is important to note that this does not mean that high-tech industries, e.g. the pharmaceutical industry, are less driven by market or managerial knowledge than low-tech industries, e.g. the pulp and paper industry. It means rather that the paper producer may be more efficient due to rationalisation and process innovation than due to product innovations.

industries. In a synthetic knowledge base, learning through interaction within the professional community (interactive learning) is dominated by tacit knowledge due to the importance of the "learning-by-doing" experience and practical skills required for the innovations (typically incremental).

In a symbolic knowledge base, innovations occur through interaction with clients as well as the overall society and professional community. Symbolic industries, such as media or entertainment, draw heavily on creative production and cultural awareness that is strongly embedded in the local community context. Therefore, knowledge flows and networks are expected to be primarily locally configured, and firms rely on informal knowledge sources rather than on scientific knowledge or principles. The role of a scientific knowledge base in the arts and humanities interplays with market and social trends. The role of tacit knowledge is great due to more concrete "know-how," craft, and practical skills. Symbolic knowledge leads mainly to incremental innovation.

Figure 1.2 suggests that all types of knowledge and all actors involved in the process of knowledge creation interact with each other. Innovation is a result of causal relationships and associated institutional innovations[4] that strengthen research-development linkages and shape the roles of the actors involved in the process of knowledge creation.

The general trend towards the complex technological knowledge required for the process of innovations, especially in high-tech products and processes, typically combines many types of knowledge from several scientific disciplines (Pavitt 1998; Johnson et al. 2002a). That is, as Gibbons, et al. (1994) put it, "the process of knowledge production is moving from the traditional disciplinary towards a new model marked by transdisciplinarity and heterogeneity." The fact that some of the actors are better qualified to generate specific types of knowledge explains their growing importance in the interplay of different knowledge sources in the creation and innovation processes. This is in line with Zahra and George (2002), who believe that knowledge in a specific industrial field is essential for radical innovation but may fail in the midst of it implementation because of a lack of sufficient expertise to resolve complex or unusual problems (Katz and Du Preez 2008). Similarly, Tripsas and Gavetti (2000) warned that deep knowledge in a specialized field may generate cognitive inertia, which constrains the firm to its current market segment or established technology for minor improvement (Levinthal and March 1993) but deteriorates its ability to pioneer using emerging technologies (Christensen and Bower 1996).

---

[4] According to Vernon Ruttan (2006, pp. 252–172) institutions are the rules of a society or of organizations that facilitate coordination among people by helping them form expectations, which each person can reasonably hold in dealing with others. Institutions, like technology, must change if development is to occur. Ruttan also considers cultural endowment, religion, and ideology as important factors of institutional innovation.

1.2.3 Knowledge Absorption, Diffusion, and "Technological Proximity"

As was pointed out in Sect. 1.1.1, the role of new scientific or technological knowledge (technical information and know-how) in technological change depends on the rate of its diffusion across economic actors and institutions. Many studies emphasize the importance of the dissemination and adoption of technical information and know-how for technological change (Denison 1979; Griliches 1980; Kendrick 1981). Technology can diffuse in multiple ways between different industries, across time and over space. The delay of the diffusion of a new technology[5] is related to the proprietary nature of intellectual property rights (IPR), to the lack of information about the new scientific or technological knowledge, or the poor capacity of regional agents to absorb it (Soete and Turner 1984). Although modern communication and broadband infrastructure can facilitate information flows, it is also essential for the development of lifelong learning, which is embedded within the knowledge base of a particular technological area of study.

Technology also diffuses through the transfer and mobility of skilled labor, the activities of professional societies and trade-investment relationships. The mobility of workers, namely highly skilled ones, is considered one of the most influential channels for knowledge transmission (Saxenian 1994). Finally, diffused technologies are acquired through scientific and social networking, including interactions among individuals from various public technology centers, government laboratories, universities, and companies. Although R&D collaboration networks offer promising and indispensable opportunities, there seems to be little consensus between researchers on the importance of networks in innovation in environments of complex technologies and rapid technological advances (Powell et al. 1996; Gertler and Levitte 2003; Porter et al. 2005). The specific nature of knowledge interactions in networks and the role of spatial proximity for the spillovers of each type of knowledge are highly disputed. While some authors have argued that knowledge networks play a pivotal role (Storper 1997; Feldman 2001), others have noted that technological knowledge is developed and exchanged in formal, more closed environments, e.g. in biotechnology (Audretsch and Stephan 1996; Zucker et al. 1998; Powell 1998; Swann et al. 1998; Feldman 2001; Lawton-Smith and Bagchi-Sen 2004). Thus, despite the ever-growing body of literature on this topic, the significance of knowledge networks remains unresolved.

One reason for the poor R&D networks' efficiency in creating new knowledge is the difference in the absorptive and diffusive capacities between regional actors. The capacity of public and private R&D collaborations and knowledge spillovers to integrate inflowing knowledge and contribute to further innovations depends on the ability to absorb and adopt the new knowledge. The latter, as explained in Sect. 1.1.1, is shaped by the level of domestic R&D and a stock of accumulated

[5] Technology includes "hard" technologies (such as computer-controlled machine tools) and "soft" technologies (e.g. improved manufacturing, quality, or training methods).

knowledge that can be embodied in local human capital (their qualifications and technical competence). Cohen and Levinthal (1989) showed that firms need to substantially invest in R&D activities in order to uderstand and evaluate new technological trends and innovations. Furthermore, the process of integration and application of knowledge cannot take place without qualified labor or human capital. Abramovitz (1986), Capron (2002), and Florida (2002) emphasize that a significant part of technology is formed by the body of knowledge acquired by people with access to higher levels of education or to training programs. In other words, domestic innovation must already have reached a critical level before foreign technology can be successfully adopted. Jindra and Rojec (2011), Rojec and Knell (2007), Rugraff (2008), and Sinani and Meyer (2009) analyzed the benefits of FDI knowledge flows for the CEE countries and noticed that, despite the friendly business environment, multinational enterprises (MNEs) did not have sufficient incentives to interact with the local R&D environment. They attributed this to a lack of absorptive capacity in domestic firms.

Furthermore, even though absorption capacity increases the region's ability to welcome new knowledge, it does not guarantee the successful adoption and diffusion of the new technology. In their works, Cowan and Foray (1997) and von Hippel (1998) implied that knowledge is not freely available to the entire population once it has been created (Jaffe et al. 1993). The studies suggest that, besides a well-developed university and R&D infrastructure, it is assumed that technological proximity between regional agents influences innovative activity. In other words, the capacity to take productive advantage of another firm's stock of knowledge depends heavily on the extent of technological similarity (Cowan and Foray 1997; von Hippel 1998; Kremer et al. 2001). The reason for this is that each technology embodies a kind of unique language and concerns a specific set of applications (Greunz 2003). Studies by Kremer et al. (2001) show that diffusion of new knowledge clearly increases through agents acquiring some existing knowledge relative to their field of studies. The reason for this is that knowledge diffusion takes place when an agent broadcasts his knowledge to the agents with whom he is directly connected. The inflowing knowledge can be beneficial only to those agents who are at least partly capable of understanding and integrating it, so as to stay within the knowledge network (Cowan and Jonard 2003). As a result, an agent receives knowledge or information he is able to integrate with his existing stock. Since a region's productive system is nothing more than a set of firms acting in different technologies, spillovers are expected to be higher between regions with similar technological profiles (Greunz 2003; Wintjes and Hollanders 2011).

In summary, technological change is a process of invention, innovation, and diffusion of technological knowledge. The innovation process is determined by the accumulated knowledge (research findings and learning-by-doing experience), the institutions that support knowledge absorption and diffusion processes (universities and R&D units), and the dynamic interplay of people within organizations and among them. The rapidly growing and increasingly complex research and innovation processes require a large variety of knowledge types and sources; yet, the degree of transdisciplinarity and heterogeneity may differ from industry to industry. New products in the field of biotechnology or ICT typically combine many technologies from several scientific disciplines. The capacity of regional agents for R&D collaborations and knowledge creation depends on their absorptive and diffusive capacities as well as the extent of technological similarity between the regional agents in the knowledge network.

1.2.4 Role of Space in Knowledge Spillovers and Diffusion Processes

The process of creation and diffusion of technological knowledge also depends on the efficiency of the territorial organization in knowledge production, which can facilitate knowledge spillovers and further advance the diffusion and development of technology. There is a rich literature giving evidence of the impact of space on nowledge spillovers[6] (e.g. Surinach et al. 2007; Decarolis and Deeds 1999; McEvily and Zaheer 1999; Almeida and Kogut 1999). One widespread assumption among academics studying knowledge spillovers has been, and largely still is, that knowledge sourcing tends to be facilitated by short geographical distance between the learning counterparts. Most empirical studies find evidence of localized knowledge spillovers (LKS) (Gertler 2003; Storper 1997; Storper and Venables 2004; Breschi and Lissoni 2001). This body of literature estimates a technology creation function and illustrates the importance of the physical proximity of firms for knowledge spillovers. The authors argue that agglomeration and population density encourage knowledge flows between various actors involved in the process of innovation (universities, businesses, and the public sector). These actors can be located in close geographic proximity (intra-regional or locally), somewhere in the country or neighboring countries (inter-regional), or elsewhere (globally).

The phenomenon of the LKSs can be explained by the fact that information about novelties flows more easily among agents within the same area, thanks to social bonds and frequent face-to-face contact. Therefore, innovations diffuse faster between individuals or organizations from the same locality (Breschi and Lissoni 2001). This is especially true for industries in which the research or innovation process requires intensive personal interactions and might favor local or intra-regional levels over others. Large literature on clusters tends to emphasize the significance of local knowledge circulation and collective learning as a source of innovation and dynamic growth (Swann et al. 1998; Feldman 2001; Lawton-Smith and Bagchi-Sen 2004; Zucker et al. 1998; Keeble and Wilkinson 2000; Feldman 2001).

Furthermore, a number of studies have found that the propensity for innovative activity to cluster spatially depends on the types of knowledge. As Audretsch (1998,

[6] "Knowledge spillovers" refer to the unintentional flows of knowledge to others beyond the intended boundary. If knowledge is exchanged with the intended people or organizations, it is a "knowledge transfer"; any knowledge that is exchanged outside the intended boundary is a spillover (Verspagen, 2008).

p. 23) suggests, the LKS and propensity for innovative activity to cluster spatially "will be the greatest in industries where tacit knowledge plays an important role (...) it is tacit knowledge, as opposed to information, which can only be transmitted informally and typically demands direct and repeated contacts." As Feldman (1994) argues, innovation-driven industries, in which tacit knowledge and innovative activity play an important role, show a higher tendency to spatially cluster. However, there have been only a few attempts to empirically investigate the role of spatial proximity for explicit knowledge spillovers, because from the methodological point of view, tacit knowledge flows are hard to track, especially when pure technological externalities are concerned (Johansson 2005). On the other hand, innovation plays a larger role in knowledge-intensive industries and high-tech industries, such as biotechnology, as well as industries that are undergoing rapid technological change or are in a growing stage of their economic life cycle. More recent developments in the literature of knowledge spillovers bring important evidence to this observation. Jaffe and Trajtenberg (1996) and Johnson et al. (2002b) record the application year of cited patents and investigate changes in the locations of citing patents over time. They find that, in the early years after the patent is approved, citations are made disproportionately by inventors in the same country (a localization effect). However, in later years, the proportion of foreign citations increases (an internalization effect). This suggests that knowledge becomes available to people across long distances only after some time, thereby confirming the benefits of proximity in the production and timely acquisition of cutting-edge technology knowledge. A similar conclusion could be drawn from results of Patel and Pavitt (1991), which show that multinational firms are strong in patenting in the areas in which their home national economies are generally strong in innovation, showing strong dependencies between the organizational capacities of multinational firms and the national systems of innovation of their home countries.

Bearing in mind the studies of Jaffe and Trajtenberg (1996), Johnson et al. (2002b), and Patel and Pavitt (1991), one could assume that the role of geographical proximity in knowledge spillovers is further determined by the importance of scientific, technological, market-related, and managerial types of knowledge in the process of innovation. Despite the fact that technological knowledge has become significantly more global in recent decades, knowledge associated with new scientific discoveries can have a high tacit, and therefore local, component (Zucker et al. 1998). Thus, scientific entrepreneurs are likely to have some advantages in identifying and exploiting new business opportunities. Likewise, relationships with research organizations, namely those conducting frontier research, can be critical for their development (Murray 2004; Bagchi-Sen 2007; Witt and Zellner 2007). On the other hand, the growing variety of fields necessary for biotechnology development suggests a more distributed nature of knowledge production; firms and R&D units may need to resort to a variety of organizations in diverse locations (McKelvey et al. 2003; Owen-Smith and Powell 2004). Managerial and organizational knowledge may come from many different locations, whereas market-related knowledge may have local or broader character, depending on the specific target markets (existing and potential customers, expectations, preferences, etc.) Unfortunately, any closer identification of the role geographical proximity plays for specific types of knowledge and technological fields is limited by a lack of empirical evidence.

1.2.4.1 Knowledge Spillovers and Urban Agglomeration

Knowledge spreads more rapidly in major cities or urban areas. The interactive, huge, and diverse nature of social capital makes large agglomeration regions an ideal space in which social learning and innovation processes can unfold (Wolfe and Bramwell 2008; Orlando and Verba 2005; Moomaw 1981). Using a large survey of Internet usage among US firms, Forman et al. (2005) show that, on average, firms in large cities adopt Internet technology faster than those in small cities or rural areas. Cities may also offer diverse human capital, which, according to Florida, is ideal for nurturing creativity and innovation (2001).[7] Wolfe and Bramwell (2008), claim that larger cities with "their mix of different industries and occupations, increase the potential for knowledge flows between industries and, therefore, exhibit faster growth and higher levels of innovative dynamism." Hagerstrand (1952) named this observation a "hierarchical effect" and "neighboring effect." The neighboring effect refers to the diffusion of technology, which happens mainly through face-to-face contacts of geographically proximate people, whereas hierarchical effect refers to the process of technology diffusion from large urban areas to smaller cities. According to Hagerstrand, a large city is a producer of new technology, while smaller cities are technological adapters. New products are usually developed and produced in urban centers, where the inventors have better access to consumers' tastes. Early production should also be in urban centers because it requires constant feedback from consumers and producers of intermediate products. As the technology matures and becomes standardized, production can move to suburban and rural areas, where cheaper labor and cheaper land is available. The studies of Storper and Scott (1987) and Storper and Walker (1989) analyze the long-term shifts of high-tech centers. They argue that the center of new technology is likely to be outside the old center, but once set, the new center will prosper until another round of new technology comes up. Again, they suggest a possible hierarchical effect in technology diffusion.

On the contrary, Rigby and Essletzbichler (1997) suggest that the technological position of states is "path dependent" and tends not to change over time. In addition, the evolutionary interpretation of technological support works by Saxenian (1994), Storper (1997), and Storper and Salais (1997), shows that regions have specific entrepreneurial cultures. Studies on the ICT and computer networks show that development of these industries reinforces concentrations of high value-added producer

[7] Florida examines the relationship between measures of diversity and tolerance and high-technology success in the 50 most populated metropolitan areas in the United States. He shows the relationship between high concentrations of creative people or "bohemians" and high technology success. One of the measures of diversity and tolerance is the gay population's rate in a metropolitan area (2001).

services in a few large metropolitan centers (Castells 2001, pp. 222–231). Similarly, Zook (2005) shows that Internet-related production of goods and services is not distributed according to population patterns but according to geographic concentration of the information economy. Notwithstanding, the importance of hierarchical models and the evidence of other mentioned studies on the role of urban agglomerations in explaining technology diffusion remains unsolved, and it is still unclear why metropolitan locations of similar sizes have different technological advantages.

1.2.4.2 Universities and Localized Knowledge Spillovers

The geographic concentration of production suggests a greater propensity for innovative activity to cluster spatially in industries in which R&D, university research, and skilled labor are important inputs. Research collaborations between universities and industry are considered to be an important channel of potential localized knowledge transfer and spillovers. Educational institutions are expected to have a positive effect because universities act as platforms for local knowledge network creation. Most of the knowledge universities produce may flow and spill over to the local economy by means of university-industry transfer projects, university spin-offs, and the mobility of university graduates and researchers to industry and social networks. Despite the importance of university-industry knowledge flows for local knowledge diffusion, studies based on systematic and longitudinal datasets remain very limited in the EU context and almost nonexistent in regards to university spin-offs, mobility, and social networks.

The pioneering work of Jaffe (1989) found that, for each U.S state, the number of patents held by businesses is positively correlated with the R&D expenditure of universities located in the same state. Based on this finding, Jaffe argues that there may be a knowledge spillover from universities to businesses. Similar results were obtained by Acs et al. (1992), who applied and enriched Jaffe's methodology with different data. Jaffe et al. (1993), using patent citations, managed to track knowledge flows from academic research to corporate R&D. The results show that firms are more likely to quote research from a co-localized university that conducts relevant research than from similar universities located elsewhere. Almeida and Kogut (1997) conducted an analogous research for the semiconductor industry. They used patent citations to show that knowledge spillovers from universities to firms are highly localized. More recent findings for the EU will be discussed in the section of Chap. 4.

In terms of spin-offs, most of the studies (Autio and Yli-Renko 1998; Stankiewicz 1994; Mustar 1995, 1997) show that spin-offs activity in EU countries is significantly lower from the US experience. In addition, the authors report that European academic scientists traditionally wanted to be involved only in research and that existing spin-off companies did not possess any long-term growth strategy (Franzoni and Lissoni 2009).

In general, the university-industry knowledge flows in terms of spin-offs are especially high in IT, biotechnology, and the medical sectors.

Another form of university-industry knowledge flows is the mobility of university graduates and researchers to industry. Trained science and technology (S&T) graduates may look for their first jobs in an area of the university (Jaffe 1989; Bania et al. 1992). Other studies in this field include the works of Bekkers and Freitas (2008); Zucker et al. (2002); Murray (2004); Edler et al. (2011); Mangematin (2000); and Gaughan and Robin (2004).

Bekkers and Freitas (2008) tackled the analysis of labor mobility (PhDs and academic staff) from the perspective of the transfer of academic technological 'breakthroughs' into the biotechnology industry in Dutch universities. The authors conclude that labor mobility is perceived to be very important in such activities. Further on, Zucker et al. (2002) report that biotechnology firms that collaborate with 'star' scientists are more likely to be productive in terms of number of patents.

Mangematin (2000), Gaughan and Robin (2004), Stephan et al. (2007), Bekkers and Freitas (2008) and Robin and Cahuzac (2003) investigate the role of S&E doctoral graduates in knowledge transfer to industry. They conclude that the careers of many S&E doctoral graduates develop across academia and industry. In their study of various channels of knowledge transfer, Bekkers and Freitas (2008) show that the importance of mobility of doctoral graduates to industry is higher than the importance of university spin-offs, joint academic-industry research contracts or industrial consultancy by academic staff. Meanwhile, Stephan et al. (2007) point out two mutually reinforcing ways of knowledge flows between doctoral S&T graduates and industry. On the one hand, doctoral S&T graduates of pharmaceutical or engineering industries employ their academic knowledge in industry; on the other hand, they learn from their training in laboratories in large corporations.

Finally, universities may actually contribute to the local economy by social interactions between firms and university faculty in a region. Breschi and Lissoni (2001) argue that knowledge does not automatically spill over into a local economy but is diffused in communities of practitioners through social networks. Similarly, Murray (2004) argues that academic inventors bring not only their technological knowledge, but also their social capital contacts, which enable companies to build networks with other scientists and research laboratories. This is supported in a study by Formahl et al. (2005), which emphasizes the role of graduates in knowledge dissemination in a more informal way, e.g. through public meetings, conferences, consulting, or information exchange. Unfortunately, there is very limited evidence for the role of social interactions in university-industry local knowledge spillovers, which leaves a question mark behind the validity of spillovers from universities and their impact on local universities.

1.3 Scientific Collaboration and Knowledge Networks

With the expansion of multinational firms and global technology sourcing, national and European research policies have focused on building networks of distributed R&D around the EU to help promote local knowledge and create sources for new technology development. "Knowledge networks" are regarded as the emerging organizational mode in environments of global competition, complex technologies, and rapid technological advances. Nonaka and Takeuchi (1995) and Lundvall and Borrás (1998) suggest that the process of knowledge generation and exploitation requires a dynamic interplay and transformation of tacit and codified forms of knowledge as well as a strong interaction of people within organizations and among them. Thus, the knowledge creative process becomes increasingly inserted into various forms of networks and innovation systems (at regional, national, and international levels). Scientific collaboration and knowledge networks can help participating entities to address the problems of complexity and rapid change in their business environment. They do so by enabling and enhancing inter-organizational learning by accelerating and supporting the diffusion of new technological knowledge (Protogerou 2010). Secondly, they help promote access to complementary resources, which is an important requirement for the creation of complex technological systems that involve a large number of different knowledge fields, which is necessary in such complex technologies as computers, semi-conductors, telecommunications equipment, and biotechnology.

Relatively few empirical studies have investigated the EU spatial dimension of R&D networks. This may be due to the fact that the majority of R&D networks between individual researchers and R&D labs have had an informal character. In recent decades, these collaborative networks have become more formalised and, therefore, it has become more possible to track them due to EU-funded Framework Programs (FP) that have allowed for more detailed analysis of networks' topology, their evolution over time, and their efficiency as mechanisms of knowledge diffusion. The formation of these networks is central to the process that should lead to the European Research Area (ERA).[8] The creation of ERA is seen as a critical step towards the development of a broad infrastructure for scientific and technological development, i.e. a set of interconnected elements that support research and diffusion in Europe and facilitate the circulation of information and knowledge, the development of transnational organizational forms, the definition of common standards, and the promotion of shared values. Technology collaboration networks formed in the context of the European FP are used as tools to re-establish private investment

[8] The main goal of the ERA is to strengthen networking between research partners and across research disciplines, as well as to link geographically-distant centers of excellence and to disseminate knowledge across Europe. Data on research networks financed by the EU within different FP are publicly available through the CORDIS.

incentives and risk and cost sharing, uncertainty reduction, and knowledge spillovers internalization.

The 4th, 5th, and 6th of the EU Framework Programmes (Peterson and Sharp 1998) were the main policy tools. Several studies started to investigate the effectiveness of the FP in building such an infrastructure (e.g. Caloghirou et al. 2004; Breschi and Cusmano 2004; Roediger-Schluga and Barber 2006). Breschi and Cusmano (2004) and Roediger-Schluga and Barber (2006) focused on R&D networks promoted under the first five FPs, and showed that the FP funding schemes support the construction of complex networks with structural properties that facilitate dissemination. Moreover, it is shown that there is a significant overlap of participants for consecutive FP and recurring patterns of collaboration amongst the same organizations that already have high R&D potential (Roediger-Schluga and Barber 2006, p. 36; Olechnicka 2012, p. 207).

In an assessment of Information Society Technologies (IST) in FP6, Breschi et al. (2007) found that the examined IST research programmes played an important role in generating and diffusing knowledge, as they managed to attract key industry players and boosted network connectivity. The key issue, which has been much less analyzed, refers to how these research networks impact regional systems and how they interact with research and diffusion activities carried out at the regional level. However, as Storper (1997) emphasizes, one of the most relevant issues in a knowledge-based economy is the tension between globalization and "territorialization." In other words, the success of the EU-supported R&D networks lies within the complementarity between local and distant interaction in order to foster an effective process of knowledge creation and dissemination (Bathelt et al. 2004). The studies show that in-house R&D seems to create a precondition allowing firms and R&D institutions to enjoy complementarity effects (Stuart et al. 1999; Caloghirou et al. 2003). For example, a study by Caloghirou et al. (2003) shows that EU-funded technology collaboration networks' success depends heavily on the closeness of the cooperative research to the in-house R&D efforts of the partner, to the firm's willingness to learn from other partners, and to the absence of knowledge appropriation problems between partners.

Summing up, the geographical dimension plays an important role in the process of creation and diffusion of technological knowledge. The process of innovation requires interaction and complementarity between local and distant sources of knowledge. Universities act as platforms for local knowledge network creation; however, in more complex technologies and rapid technological advances, it is essential to promote access to complementary resources of knowledge through technology collaboration networks. Yet, the success of the integration of technological knowledge, inflowing from elsewhere, depends on the network structure among socio-economic agents within local economies. It spreads more rapidly in major cities and large agglomerations.

1.4 Innovative Potential, Competition, and Measures of Technological Change

1.4.1 Technological Change and Growth of High-Tech Industries

The processes of liberalization and globalization also affect patterns of knowledge creation and diffusion. In practice, globalization and technological advancement are mutually reinforcing processes. On the one hand, globalization and international economic integration offer an important channel for flows of foreign technology that embody significant innovations, enhance personal mobility, and push competitive pressures on local firms. Furthermore, technical and technological advances, notably in the transport and communication industries, lower costs and foster international trade, investments, and mobility of people around the world.

Technology transfer through multinational enterprises (TNC) has become an increasingly important driver of both globalization and technological change. Trade and foreign direct investment flows (FDI) provide domestic firms with technological, market, and managerial knowledge, and they induce more R&D. Transnational enterprises (TNCs), with their foreign subsidiaries, can act as demonstrators for local producers or can directly help local firms by providing training for technicians, engineers, and managers. Imported goods may serve as channels of knowledge spillovers because of technological innovations embodied in these goods. In this way, liberal trade and FDI flows enrich the regional knowledge stock (Coe and Helpman 1995; Keller 1997; Porter, 1990).

On the other hand, the technological advancement in ICT made it possible to manage the value chain and perform activities in any location in the world. It enabled the largest latecomer firms to improve their competitive advantage by developing new and advanced products and processes along with foreign companies. In these joint ventures, local companies usually provided advanced know-how of manufacturing processes, while foreign partners provided financial capital and access to their basic research facilities at home.

Technological advancement and globalization create opportunities for regions and locations to upgrade their technological and industrial capabilities and to integrate into the global value chains (GVCs). Furthermore, the route of integrating in the global economy exposes a host country's macroeconomic and business conditions that heighten the competitive pressures, stimulating regions to make better physical infrastructure and utilities, to create a more business-friendly environment to invest in knowledge and R&D, to upgrade industrial capabilities, and to move up the value chain. For example, firms in Taiwan Province of China moved from the manufacture of transistor radios to calculators, to TVs, to computer monitors, to laptops, and more recently to Wireless Application Protocol phones (WAP) (UNIDO 2004, p. 10). Furthermore, the outsourcing and offshoring of less efficient activities to other locations increases productivity of home companies.

A globally competitive market may force firms and industries to become more engaged in innovative activities and to improve their productivity or move into knowledge-intensive and higher value chains, e.g. into high-tech sectors. Exactly for this reason, high technology industries have been more internationalised or affected by GVC. According to the Organization for Economic Co-operation and Development (OECD), high-technology products include technical products, the manufacturing of which requires high R&D intensity (Hatzichronoglou 1997). This definition implies an important linkage between high-tech manufacturing and innovation activity. High-tech manufacturing requires almost twice as many technical personnel and devotes twice as many resources to scientific R&D (Moriarty and Kosnik 1989). Furthermore, the development and introduction of new high-tech products and/or manufacturing processes requires the systematic application of scientific and technical knowledge (according to Hecker 2005).

The pressure to move the regions' and locations' specialization towards more knowledge-intensive activities is also reflected in trade flows. In fact, trade in high- and medium high-tech industries has grown faster than total manufacturing trade in the OECD area. High-technology industries are the most dynamic manufacturing industries, representing about one-quarter of total OECD trade (OECD 2007). A similar trend can be observed in knowledge-intensive services (KIS).[9] Knowledge-intensive services, especially high-tech knowledge-intensive services (HKIS),[10] which consist mainly of R&D and ICT industry-related services, play a vital role in innovations and technological change processes in several ways. HKIS contribute interactively to the process of knowledge creation and innovation. In this context, they promote learning processes in firms and R&D organizations, facilitate knowledge flows from one sector to another, and may further be considered as "bridges for innovation" (Antonelli et al. 2000; Czarnitzki and Spielkamp 2000, p. 26). A study by Makun and McPherson (1997) indicated that innovation rates are greater in locations where a high concentration of HKIS exists. Furthermore, the development of HKIS, Internet, and communication in particular, contributed to the development of outsourcing and partnering in R&D. Outsourcing in many high-tech industries enabled restructuring and rapid organizational growth, and therefore accelerated technological advance. High-tech companies now outsource lots of professional jobs offshore. These jobs include engineering, prototyping, testing, consulting, statistical analysis, and many others. For example, offshoring in the IT industry is no longer restricted to basic service tasks such as data entry and processing, but has expanded to include sophisticated works such as knowledge services, decision analysis, design, engineering, and R&D.

---

[9] Following the aggregations of services based on NACE Rev 1.1, the KIS include: 61 – Water transport; 62 – Air transport; 64 – Post and telecommunications; 65–67 – Financial intermediation; 70–74 – Real estate, renting, and business activities; 80 – Education; 85 – Health and social work; 92 – Recreational, cultural, and sporting activities. Data from October 2011, Eurostat.

[10] Knowledge-intensive high-tech services: Post and Telecommunications (64); Computer and related activities (72); Research and development (73), Eurostat, Statistics in Focus, Science and Technology, 4/2005, R&D Statistics, Luxembourg, 2005.

The extent of offshoring and outsourcing may be hard to measure; however, the existing literature discussing both challenges and opportunities indicates immense advantages of offshoring and outsourcing, both to the outsourcer and outsourcee (Barney 1991; Walker et al. 2005; Hoecht and Trott 2006; Langlois 1990; Strambach 1998). One of these advantages is that companies can contract some of their technological processes to companies or R&D institutions that have better resources and expertise in specific fields.[11] The latter allows saving the company's cost as investing in new technology. Furthermore, companies' productivity can be increased by employing skilled manpower in larger numbers at lower costs. The negative attitudes toward offshore outsourcing were discussed mostly by parties in the US and UK due to job losses in the mentioned countries (Lieberman 2004; Hicks 2004; Kathawala 2005). However, other authors, including Makun and McPherson (1997), showed that – despite the Internet – the major part of outsourcing is not possible because of the need to establish face-to-face contact in order to transmit tacit knowledge.

In summary, the increase of HKIS services and the possibility of offshoring, in particular, can be seen as an opportunity and a real challenge by many regions, sectors, and groups of workers. They reinforce local productive systems due to the "cumulative dynamics" of knowledge (creation, accumulation, and diffusion of knowledge). HKIS and the possibility of offshoring may be a potential compensation for low domestic R&D intensities through the importing of knowledge from elsewhere and, therefore, contributing to future technological developments. The latter is especially true for industries heavily exposed to international competition. Regions specializing in these industries may be further marginalized in the global economy unless they can upgrade their industrial sectors by gradually moving towards higher added-value production. This, on the other hand, requires upgrading skills and technological capabilities of local clusters.

1.4.2 R&D and Patents as Measures of Innovative Potential of Regions

Technological advancement is a complex phenomenon, and attempts to empirically measure it are extremely problematic. The subject literature reveals lots of confusion related to what is the genuine measure of technological change. The following section reviews the common views of each of these measures representing technological change.

[11] Many companies have set up their own R&D centers in India or China in order to reduce complexity and cost of their production processes in such sectors as IT and electronics. As a result, both of these countries became almost overnight global hubs for these industries; http://www.design-reuse.com/news/8711/outsourcing-india-shifts-d.html; http://www.chnsourcing.com/top50/2008/.

1.4.2.1 Indicators of Research and Development (R&D)

A number of studies have relied on indicators such as R&D spending or R&D intensity (Mowery and Rosenberg 1989) as a proxy for technological change. Freeman (1994) argues that patents and expenditures on R&D are at best proxies for technological change but should never be seen as truly representative. On the other hand, Wilson (2001) considers a variable that could be related to the true rate of technological advance would be the amount of R&D that went into developing the technology embodied in an industry's capital.

Furthermore, Scherer (1999) founds that a firm's R&D expenditure corresponds with its own valuation of the resulting patents, which implies that R&D that results in technological change is measured by the value of new patents. Studies by Jaffe (1986) and Trajtenberg (1990) prove that R&D in a particular technological field is proportional to technological change, measured by simple patent counts. A more recent study, conducted by Wilson (2001), supports the former research. The author found that R&D spending within particular technological or product fields is the primary driver of technological change in those fields.

The greatest weakness of the R&D indicator of technological change originates from the fact that it measures inputs to technological change rather than technological change itself. Consequently, R&D investments or intensity measure rather innovative potential that can induce faster technological advance. Another shortage of R&D measures as technological change is the fact that they do not reflect "craft and learning" – other important components of technology. This dimension is simply not the product of R&D activities. Furthermore, Patel and Pavitt (1995) argue that the "widely used R&D indicator is better at measuring technological activities in the science-based classes of technology (e.g. electronics) than in the production-based and information-based classes (e.g. software)" (p. 20). This is because R&D expenditures and R&D personnel consider only those activities that are formally conducted in R&D labs. In many high-tech industries in which technological changes occur mostly in the design or production division, such as mechanical engineering, the number of patent counts may be significantly higher than the share in R&D expenditure (Townsend et al. 1981). For this reason, the R&D activities of small firms may be seriously underestimated because of their lack of R&D infrastructure.

1.4.2.2 Patent and Patent Citations

The purpose of a patent is to provide protection for technological advances by guaranteeing intellectual property rights to the inventor. To sustain their advantage, companies must protect their innovation (it is difficult for a first mover to defend its market position against late entrants). In practice, however, large players use patents most often. Patent protection is especially needed for radical innovations. A number of company surveys conducted in the United States, Europe, and Japan in the mid-1980s and 1990s reported that patents were extremely important in protecting competitive advantage in a few industries. This was especially true for biotechnology, drugs, chemicals, and computer industries (Levin et al. 1987; Cohen et al. 2000). However, patents may also reduce a monopolist's motivation to introduce a next-generation of radical changes (Arrow 1962; Geroski 1989). As a result, the new challenger (the one that introduces the new-generation product) would receive the total benefit, while the incumbent monopolist (patent holder) would gain the difference between new-generation rent and current monopoly rent (Aghion and Howitt 1992). Patents obtained on radical innovations can be used to protect these innovations from incremental improvements made by other companies. In contrast, the current patent holder may be interested in introducing incremental innovation protected by utility models to reinforce its monopoly rents, whereas incremental improvements involve significant changes, in which case innovations can be patented.

Measuring the level of patented inventions provides an indication of the potential capacity for innovation and technological advance. Innovations that pass the rigorous patent review process are assumed to have a high degree of invention. A patent is granted only to a new and useful device. In this case, the term "useful" refers to the condition in which the subject matter has a useful purpose and also includes operativeness (Worgan and Nunn 2002). This implies that a patented invention may have a practical impact on technological advances. In addition, the patent system advertises and disseminates knowledge about new technologies (Lamoreaux and Sokoloff 1996). The patent data contains important knowledge that inspired current inventors and will inspire future inventors.

The greatest weakness of patents data as a measure of technological change is that, first of all, not all innovations are patented. Many technological changes cannot be patented. Improvements in organizational principles, learning, and craft do not necessarily have a "new device" to patent. In addition, the radical innovation required for the creation of new products and markets does not guarantee the success of the inventor. In fact, incremental innovation can influence technology in a more significant way and be more beneficial to companies than radical innovation. Large companies often do not want to take risks by investing in something precarious; however, once they recognize that a new product is likely to become popular, they have an interest in introducing their own improved version in order to gain their share of a market that is continuously growing. Once they seize the opportunity, large companies often have means to "close the window behind them" by acquiring intellectual property, locking customers in or through other strategies (Striukova and Rayana 2009).

Furthermore, many patented items remain on paper without much effect on the industry or technology (Desrochers 2002; Feldman 1994). This apparent weakness turns out to be an advantage in some cases. However, even inventions that have not been commercialized should not be overlooked because they may influence inventions of commercial value indirectly later, e.g. through learning. Learning and craft bring increasing importance to many high-technology fields such as aviation. Therefore, one cannot assume patent data captures all the dimensions of technology.

Moreover, patents may have a negative effect on competition and technology diffusion, and hamper innovations, especially when they limit access to essential knowledge, as may happen in emerging technological areas. In this context, too broad protection on basic inventions can discourage follow-on inventors, if the holder of a patent for an essential technology denies access to others under reasonable conditions. This concern has often been raised for new technologies, most recently for genetic inventions (Bar-Shalom and Cook-Deegan 2002; Nuffield Council on Bioethics 2002; OECD 2003) and software (Bessen and Maskin 2000; Bessen and Hunt 2003).

In addition, as has long been recognized, the main drawback of patents is their negative effect on diffusion and competition. As patents make an exclusive right that creates a temporary monopoly, the patent holder can set a market price higher than the competitive price and limit the total volume of sales. This negative impact on competition could be magnified if patent holders try to strengthen their position in negotiations with other firms in attempts to block access by competitors to the key technology or, inversely, to avoid being blocked by them (Shapiro 2002). Such strategic patenting seems to have developed over the past 15 years, notably in the electronics industry (Hall and Ziedonis 2001).

Nevertheless, patents can also have a positive impact on competition when they enhance market entry and firm creation. Not only is there evidence of small companies being able to assert their right in front of larger ones thanks to their patent portfolio, but patents may also be a decisive condition for entrepreneurs to obtain funds from venture capitalists (Gans et al. 2002). Moreover, patents may enhance technology diffusion. Patenting means disclosing inventions that might otherwise be kept secret. Industrial surveys show that the reluctance of firms to patent their inventions is due primarily to the fear of providing information to competitors. This was confirmed in the OECD/BIAC survey on the use and perception of patents in the business community, sent to firms in OECD countries in 2003, in which respondents indicated their intensive use of patents as a source of information (Box 2; Sheehan et al. 2003). Patents also facilitate transactions in markets for technology: they can be bought and sold as property titles or, more frequently, be subject to licensing agreements which allow the licensee to use the patented invention in return for payment of a fee or royalty (Arora et al. 2001; Vonortas 2003). Finally, governments have advanced the goal of enhancing technology diffusion to encourage universities to patent their inventions, with the objective of licensing them to businesses that will further develop and commercialize them (OECD 2003).

Another advantage of patent data as a potential measure of technological change is that they include information on time and space. From the methodological point of view, this gives patent data a tremendous advantage over other measures of technology. Furthermore, a patent record contains the addresses of both its inventors and legal owners. This address information can be used for various kinds of geographical analysis.

Geographical analysis of patent increases thanks to the availability of address information. The location of inventive activities, the geographical diffusion of technology, and the regional growth literature have made good use of patent data;

this body of literature will be reviewed in the next sections. Therefore, patents remain the best single source of statistical evidence for innovative potential, knowledge diffusion, and technological change. However, practitioners and researchers should use them with appropriate care (Worgan and Nunn 2002).

Another measure of innovative potential and proxies for the value of technological change is the number of product and process innovations and scientific papers. While the number of product and process innovations is one of the best measures of technological change, it is very costly. Product and process innovations are possible due to the survey of primary and secondary sources of innovations. This is precisely why comprehensive and continuous (using time-series studies) innovation surveys are rarely conducted. Finally, academic publications have limited potential to proxy for technological change. They measure basic research outcomes rather than real technological changes. The basic research contributes to technological change only over the long run. Furthermore, not all technological innovations are published. For example, many organizational changes, as well as craft and learning-based experiences, do not appear in academic publications.

1.4.3 Productivity and Employment in High-Technology Industries

Productivity reflects all available current technologies. Productivity rises from one period to the next through new innovations and technological change. Changes in productivity have been emphasized by growth theories as an important measure of technological change (Solow 1956; Arrow 1962; Romer 1986). Each innovation increases the productivity of each worker by improving the organizational principles and learning process. Furthermore, investment in human capital generates spillover effects that increase the productivity of both physical capital and human capital in general (Lucas 1988). Workers assimilate ideas inherited from preceding workers, which implies increasing returns from human capital. However, many examples of technological change do not reflect any improvement in productivity. An example is consumer products whose main benefit to the consumer is quality, not productivity itself. In some cases, inventions may lower the productivity of labor. Many household appliances, such as washing machines, liberate people from their domestic work, increasing the supply of labor, which consequently gives companies incentives to substitute capital with labor.

Traditionally, technical change is seen as factor-neutral. However, the evidence of technological change in the past decade suggests growing demand for more educated, more able, and more experienced people (Murphy et al. 1998; Denny and Fuss 1983; Autor et al. 1997; Berman et al. 1994). In general, these studies suggest that technological change favors skilled over unskilled labor by increasing its relative productivity. Technical advancement triggered the shift in production technology; therefore, it is now skill biased. For example, evidence of the new information technologies suggests they are complementary to skilled labor in the adoption phase. The extent to which new capital will complement skilled or unskilled labor is determined endogenously, e.g. by innovators' economic incentives, relative prices, the size of the market, and institutions.

This skill bias can be explained by various factors. In past decades, technical change has required relatively many highly skilled workers. This suggests firms have become increasingly prepared to pay for more highly skilled workers, as the costs associated with the tangible part of new investments have decreased with time (Petit and Soete 1998). Moreover, employers have increasingly replaced more expensive low-skilled workers by relatively inexpensive high-skilled workers. This indicates the returns to investment in high-level education have increased and have been accompanied by a major increase in labor skills.

Griliches (1969) proposes that this skill bias is due to the relative decline in the price of capital, while Denny and Fuss (1983) attribute the skill bias to the specific effects of technical change. Autor et al. (1997) and Berman et al. (1994) argue that highly skilled workers adapt more easily to changing technologies than their colleagues with lesser skills. Murphy et al. (1998) found that new technologies are relatively complemented with more educated labor. This is closely related to the previous assumption that machinery and new technologies harm low-skilled workers. This complementary process of mutual reinforcement between education, productivity, and knowledge-intensive employment can be explained in the following way: a highly educated and multi-skilled workforce (in diverse technological disciplines) creates innovations that may impact the growth of new knowledge-intensive industries (e.g. high-tech industries). The latter, on the other hand, may improve productivity and induce technological advancement across many other industries, and create further demand for highly skilled labor and advanced technologies. Indeed, technological change in one industry increases the productivity of innovating firms belonging to this and other industries. For example, as ICT becomes pervasive, the potential of economy-wide productivity gains shifts from high-tech manufacturing to the overall economy. Obviously, in order to consider productivity as a measure of technological change, one should accompany this measure with the dynamics of the ratio of a highly qualified labor force, employment in high-tech and knowledge-intensive industries, as well as R&D and patent intensity.

1.4.3.1 Knowledge-Intensive Services and High-Tech Manufacturing

The relationship between technological change and employment is based on *product and process* innovations. The process-related innovation driving technological change is related to productivity growth (Lee 2008). For example, computer-programming technology improves the way in which information can be referenced and accessed. As a result, the unit cost of performing these tasks decreases.

The relationship between technological change and employment is also based on *product* innovation, which refers to the birth of entirely new economic branches, in which additional jobs can be created. A large number of authors agree that product innovations have a positive impact on employment, since they open the way to the development of entirely new goods or to the radical differentiation of mature goods (Freeman et al. 1982; Katsoulacos 1986; Freeman and Soete 1987, 1994; Vivarelli and Pianta 2000; Edquist et al. 2001; Bogliacino and Pianta 2010).

Technological change driven by product innovations follows the process of Schumpeter's "creative destruction." New technologies destroy jobs in some industries while creating jobs which require different skills in different industries. For example, cars displaced old products, such as carriages, thereby leading to increases in employment of the totally new industry – the motor industry (Freeman et al. 1982; Katsoulacos 1984). In past decades, there has been growing evidence of the fact that technology contributes to the creation of high-skill jobs. Skill-biased technological change refers to the shift from low-tech to high-tech specializations of regions (Entorf and Pohlmeier 1990; Blanchflower et al. 1991; Smolny 1998). However, these results are not equally distributed across industry levels. For example, the highest growth rate was recorded in occupations that are related to ICT and involve high levels of education. Furthermore, rapid advances in ICT have created new kinds of services, "knowledge services," such as data entry and information- processing services, and research and consultancy services.

The role of knowledge-intensive sectors as indicators of the state of the knowledge economy and technological change was discussed by Cortright and Mayer (2001) and Tödtling et al. (2006). Those sectors typically include high to medium high-tech manufacturing sectors and high-technology knowledge-intensive services (HKIS).[12] Strong technology intensity in these sectors is related to the importance of "know-how" and intense innovative activity. There is broad agreement regarding the general characteristics of these sectors, which include: high involvement in innovative activity, R&D intensity, high proportion of "technology oriented" personnel, intensive use of technology, high technology content (patents, licenses, and know-how) and leading-edge products or services (Mukkala 2011).

Summing, up there are no direct measures of the rate of technological change; therefore, any attempts to measure technological change, its drivers, and its impact on economic growth must be linked to several proxies. As no single proxy is perfect, it is important to use alternative measures for the analysis. If similar results are obtained with different measures, one can be more assured of the reliability of the findings (Bartel and Sicherman 1999). All the proxies can be divided into two categories: input- and output-based measures. Productivity and employment growth variables measure technological change as the rate of change in output. Input-based proxies are measured in levels of R&D intensity and patents – all have been proved in the previous section to be good proxies for the rate of technological change.

[12] The OECD has classified the manufacturing sectors according to their R&D intensity. High-technology industries include: Aircraft and spacecraft; Office machinery and computers; Electronic and communication equipment; Pharmaceuticals, medicinal chemicals, and botanical products; Medical, precision and optical instruments, and watches and clocks. Medium-high technology includes: Electrical machinery and apparatuses; Motor vehicles, trailers, and semi-trailers, railway and tramway locomotives and rolling stock; Chemicals and chemical products; Machinery and equipment.

1.5 Knowledge Spillovers and Regional Innovation System (RIS)

1.5.1 Region's "Innovation System" Versus Cluster-Based Approaches

The translation of knowledge and innovations into technological change (growth of productivity and knowledge-intensive jobs) will be successful if it is supported by the local innovation environment (embodied in the institutional, political, and social context of the relevant regional economy). There are at least two approaches to the concept of regional innovation environment that result from either the systemic or interrelated nature of innovation. The first originates from the innovation-system approach at the national, regional, or local levels. This approach is more "systemic" in explaining the endogenous capacity of companies and regions to innovate, and focuses especially on the role of knowledge creation, absorption, and diffusion (Giuliani and Bell 2005). The innovation processes within the regional innovation system (RIS) have a systemic character because they are embedded in the local social, political, economic institutions, and cultural context. In the words of Lundvall (1992, p. 2), "[...] innovation system is constituted by elements and relationships which interact in the production, diffusion and use of new, and economically useful, knowledge."

The second approach to regional innovation – the cluster or agglomeration approach – considers the interrelated nature of innovation. Because the process of innovation and technological change takes place in space, it is bounded to the local productive system. In the light of the cluster approach, regional processes of innovations result from the interrelated activity of firms, suppliers, service providers, coordinating intermediaries, and institutions such as universities or community colleges.

The founder of the cluster theory, Michael Porter, did not say much on the difference between a cluster and RIS. Thus, both in the cluster and RIS approach, the specific local context plays an important role in the innovative output of regions. This is especially true for industries based on an analytical knowledge base. RIS can play an important role in promoting new and emerging economic activity based on industries with an analytical knowledge base, requiring close and systemic industry-university cooperation and interaction, e.g. in science parks located in proximity to knowledge-creating organizations (e.g. technical universities). Clusters involve specialization, proximity, and cooperation that lead to spillovers and synergy within RIS (Doloreux and Parto 2005).

Similarly to RIS, clusters also include public institutions, such as government education institutions and support services, with cluster boundaries defined by linkages and complementary across institutions and industries (Porter 1998). The major difference between these concepts is that RIS can stretch across several sectors in the regional economy. Clusters and RIS may co-exist in the same territory and RIS may, in fact, contain several clusters.

Indeed, clusters have a complementary role in RIS; they are closely linked and mutually reinforcing. They act as two subsystems of actors and institutions that get systematically engaged in the process of knowledge exploitation and generation (Cooke et al. 1998). The first one consists mainly of firms (displaying clustering tendencies) that make up the regional production structure and are responsible for knowledge exploitation. The second is the regional supportive infrastructure, which consists of public and private research laboratories, universities and colleges, technology transfer agencies, and vocational training organizations; this infrastructure is responsible for knowledge generation and diffusion. Both of these subsystems are equally important for creating a friendly businesses environment for innovation, technological change, and growth. Chaps. 3 and 5 will attempt to empirically test the importance of both RIS- and cluster-based approaches in explaining the innovative performance of EU regions.

1.5.2 Knowledge Absorption and Diffusion within RIS

The literature on RIS as a framework for understanding the innovation process in the regional economy has grown significantly since the early 1990s. It includes works by Cooke et al. (1996, 2002), Doloreux and Parto (2005), Asheim et al. (2003), and Giuliani and Bell (2005). Although the authors broadly agree on the importance of RIS in regional innovation processes and knowledge flows, they do not suggest any widespread definition of "innovation system." Lundvall (1992) gave a general definition of a system, according to which the system constitutes a number of elements and the relationship between these elements. The elements and relationships "interact in the production, diffusion and use of new and economically useful knowledge" (Lundvall 1992). Lundvall's definition of the system of innovation is somewhat similar to the way Freeman (1987) saw the innovation system, describing it as "the network of institutions of private and public sector, whose activities and interactions initiate, import, modify, and diffuse new technologies." Cooke conceptualized it in terms of "…interacting knowledge generation subsystems, which are composed of public and private R&D institutions, universities and colleges, technology transfer agencies, vocational training organizations, and the regional production structure" (2004, p. 3). The author emphasized that the systemic character of the relationships between these institutions refers to a "regular, two-way, interchange on matters of importance to innovation" (Cooke et al. 1997, p. 484). Nauwelaers and Reid produced another, more complex definition of RIS as "the set of economic, political and institutional relationships occurring in a given geographical area which generates a collective learning process leading to the rapid diffusion of knowledge and best practice" (Wolfe 2000). Lundvall observed similar aspects of innovation, pointing out that innovation is the result of an interactive learning process between firms and their environment, ensuring territorial and social integration within a cultural and institutional context (1992). Doloreux and Parto (2004) also emphasized the importance of cultural context, arguing that successful co-operation in innovation activity between firms and other organizations, such as universities, R&D institutes, and technology transfer agencies, depends on the local culture that supports creativity and innovations.

Summarizing, all of the aforementioned definitions refer to RIS as a set of institutions and actors of technological innovation, education, and training that are systematically involved in the generation, use, and dissemination of knowledge and innovations, and share common local cultural creativity.

The major disadvantage of these approaches to RIS is that they capture the complexity of RIS as an integrated whole without sufficiently revealing what constitutes the institutional structure and the production structure, and the interrelations that bind them together. Furthermore, the above approaches ignore the fact that current RIS is an open system, affected by external influences from other "innovation systems" (regional, national, and international sources). As Asheim and Gertler (2004) emphasize, RIS "are not sufficient on their own to remain competitive in a globalizing economy." Local firms and institutions must have access to, and be capable of, incorporating the knowledge inflowing from other innovation systems. Briefly speaking, the systemic approach may be very important to understand the innovation processes within regions and to specify desirable factors and mechanisms for promoting knowledge diffusion and innovations. However, its analysis must expand beyond RIS boundaries through the process of economic integration and globalization. Hommen and Doloreux (2004) emphasize the need to expand studies on RIS. A number of authors broadly suggest that further success of the concept of RIS lies within the interlinkages of both endogenously generated and exogenously available knowledge (Cumbers et al. 2003; Mackinnon et al. 2002; Bunnel and Coe 2001; Archibugi and Michie 1997).

1.5.3 Institutional Structure of RIS and Technology Transfer

Another advantage of the systemic approach to innovation is that the interdependencies of innovation processes could be simplified and grouped into two forces of innovation: demand and supply. The supply side of RIS consists of the institutional sources of knowledge creation, such as the institutions responsible for the training and preparation of highly qualified labor power. On the demand side of RIS are productive systems, firms, and organizations that develop and apply scientific and technological output of the supply side in creating and marketing innovative products and processes. RIS enable the knowledge and technology transfer[13] from R&D and education institutions into local productive systems.

The processes of knowledge and technology transfers encompass a very broad range of activities supporting mutually beneficial collaborations between universities, businesses, and the public sector. The Triple Helix approach, introduced by Etzkowitz and Leydesdorff (2000, p. 111), includes both collaboration and cooperation among the abovementioned institutions. Interaction across different helices helps to establish "knowledge networks." The efficiency and scope of these interactions depend on the knowledge absorption capacities of each actor involved in the network, e.g. human qualifications, intensity of R&D, etc. Cockburn and Henderson (1998) argue that effective integration and transfer of new technologies depends on the intensity of R&D performed by firms, and the capacity to tie close links with universities and government R&D institutions.

Since direct measures of knowledge transfer among these institutional sectors are not available, the R&D intensities performed by these institutions can proxy knowledge transfer within the Triple Helix model (Bessant and Rush 1995; Gadray et al. 1995; Capron 2002; Greunz 2004). This can be true, especially in the case of large firms, where knowledge and information are transferred via internal functional interaction among the R&D, production, marketing, and strategic planning departments. In other words, learning occurs through cooperation and interaction between firms and the local scientific system (between the supply and demand sides of RIS). In this way, investments in R&D and local human resources could foster knowledge transfer and technological innovation in the region (Aghion and Howitt 1992; Grossman and Helpman 1991). The efficiency of R&D funding and its impact on innovations and regional growth depends heavily on the efficiency of technology transfer and diffusion mechanisms. The importance of universities as a local supply of skilled personnel and technology transfer (i.e. R&D agreements, qualified labor pool, lifelong learning, etc.) suggests the need to discuss here their roles in the above-mentioned processes in more detail.

1.5.3.1 University-Industry Interlinkages

It has been previously mentioned in this book that the absorption capacity of regions depends on the level of qualification and competence of the workforce. The higher the quality of human capital is, the higher the productivity of R&D research and the greater

---

[13] Traditionally, technology transfer included the transfer of hardware objects. However, due to the rapid development of ICT infrastructure, transfer often involves information (e.g. a computer software program or a new idea). The Work Regulations of the United Nations define technology transfer as a systematic transfer of knowledge for the manufacture of a product or provision of service (Yu 1991). Knowledge transfer is the fundamental element of technology transfer; without knowledge transfer, technology transfer does not take place. It includes the transfer of tangible and intellectual property, expertise, learning, and skills between academia and the non-academic community.

the efficiency of RIS will be. Universities and academic institutions are among the key components of the region's technological infrastructure. Traditionally, the roles of universities and colleges were to educate students and to conduct basic research. Their major contribution was to build the absorptive capacity of regions by providing a high standard of education. Over the years and throughout the scientific revolutions, universities took on another role, becoming central players in states' and regions' innovation infrastructure and economic growth. Mansfield (1991) noted that university R&D enhances the stock of basic knowledge, generates increasing technological opportunities across a wide range of industrial fields, and increases the potential productivity of private industrial R&D. Most of the literature reviews this process as a new task for universities and their relationship with government policies and businesses, and emphasizes the role of the state in changing academic science, particularly in specific government policies to encourage market-oriented research (Minshall et al. 2004; Slaughter and Leslie 1997; Slaughter and Rhoades 2004).

The university provides knowledge with which industry can develop new technologies (Breznitz and Anderson 2006; Di Gregorio and Shane 2003; Henderson 2006), while advances in technological knowledge have to be absorbed by firms and applied to the production process and organization of work. Therefore, the economic contribution of academic institutions depends on the effectiveness of technology and knowledge transfers into the private business sector.[14] In studies by Jaffe (1989), Acs et al. (1992, 1994), and Zucker et al. (1994), the authors provide evidence of positive university-industry knowledge spillovers. Jaffe (1989) and Acs et al. (1992, 1994) found that knowledge created in university laboratories spills over to contribute to the generation of commercial innovations in the private sector, whereas Zucker et al. (1994, p. 1) provide considerable evidence suggesting that the timing and location of new biotechnology firms is "primarily explained by the presence at a particular time and place of scientists who are actively contributing to the basic science." Furthermore, they emphasize that most formal and informal mechanisms of technology transfer and diffusion depend heavily on spatial proximity. This is even more the case of spillovers of tacit knowledge.

Furthermore, a cross-fertilization of R&D efforts is essential for fruitful scientific effort. Cross-fertilization of R&D efforts among different scientific disciplines provides further evidence for why innovating firms require R&D co-operation with

[14] Varga (1989) identifies several ways of university-business knowledge transfers: formal co-operations and agreements on R&D, industry financed university research centers, faculty consulting in industry, scholarly journal publications and industrial associates programs. Other forms of knowledge transfers are industrial incubators and industrial parks, aimed mainly at providing facilities to start-up firms, as well as university spin-offs.

different partners.[15] For instance, the need for basic research requires cooperation with public science institutions, e.g. universities. Collaboration with universities often contributes to radical innovation. This is because university labs most often have different R&D portfolios than companies. Accessing dissimilar knowledge through R&D collaborations decreases the rate of incremental innovation but increases the generation of radical new products. Besides, the greater product and technology complexity increases costs and risks for innovators, so that these can hardly be dealt with alone. Sharing R&D costs and benefiting from resource pooling is another important motive for cooperation between universities and firms (Miotti and Sachwald 2003). As a result, networking or co-operating with different partners on innovative activities becomes beneficial (Hagendoorn 2002).

Local technology transfer institutions can act as intermediaries and play a stimulating role in interactions between actors of the Triple Helix model, which induces more rapid adoption of new techniques (Galaskiewicz 1985; Suchman 1994; McEvily and Zaheer 1999). Such intermediaries include organizations that are either internal or external to university environments. The first group of organizations includes university technology transfer offices (TTO), the second group involves intermediate research organizations that operate as bridges between universities and firms but are autonomous, independent, and funded through combinations of public and private resources.

The study evidence of TTO's efficiency is quite mixed. Studies of US universities have shown a positive effect of TTO's experience on efficiency in producing licenses (Link and Siegel 2005) and licensing income (Link and Siegel 2005) whereas studies of UK universities found no effect (Lockett and Wright 2005; Markman et al. 2005) or even decreased efficiency in producing licenses. Studies of other EU universities emphasize the role of TTOs as facilitators of university-industry knowledge transfer in the EU (Debackere and Veugelers 2005; Geuna and Rossi 2011). Among the studies showing the limited role of TTO in technology bridging between universities and firms are works by Kienbaum (2006) and Jensen and Thursby (2001). US TTOs tend to be generally older than UK TTOs and thus these seemingly contradictory findings might result from the fact that the positive effect of TTO's experience starts only after some substantial experience has been gained.

In terms of the second group of intermediary organizations, the experience is somewhat similar. While some researchers found evidence of a positive impact (Debackere and Veugelers 2005; Sakakibara and Branstetter 2003; Sapsed et al. 2007), others found no impact or even some negative impact (Muscio 2010;

[15] Zand and van Beers (2008) emphasize that radical innovation performance is positively influenced by collaboration schemes with all partners except for competitors. The threat of leakage of sensitive information towards competitors can explain why R&D co-operation with competitors affects radical innovation performance negatively.

Wallsten 2000). For example, a survey by Shohert and Prevezer (1996) emphasizes the important role intermediaries play in helping to formalize informal collaborations in terms of contractual and licensing arrangements within biotechnology in the UK; meanwhile, in a survey run by Kienbaum (2006, p. 43), about half of the German inventors said they would have commercialized their inventions successfully even without the intermediary. At the same time, they emphasized the importance of integrating scientists into the transfer process in order to commercialize technology successfully.

1.5.3.2 Role of Government in Technology Transfer

Besides universities and the business sector, the third component of a region's technological infrastructure is the government. Government R&D is partly realized in universities, but also in national laboratories. The economic rationale behind government research is not only to satisfy public needs, but also to counterplay marked failures in the field of R&D investment as well as reduce the private cost of R&D, strengthening the capacity of knowledge creation of universities in a given strategic field, helping the private business sector to increase its absorption capacity, encouraging universities and businesses actors to actively join global international research networks, or stimulating cooperation between industry and university.

The most common government involvement in R&D support is probably support for basic research. Government may also stimulate R&D performed by business by reducing the private cost of R&D (e.g. grants) or by helping firms understand the technological opportunities available (Guellec and Van Pottelsberghe 2003). It is generally accepted that R&D and knowledge are not common private goods that are traded according to the market mechanism because of their public good characteristics. It is not always possible for firms to appropriate all the returns that result from their R&D activities, so that other economic agents are also able to reap benefits without bearing large costs. The private business sector may be discouraged from performing R&D from which it cannot capture the entire return (especially when the associated risk is high). Since the private rate of return is smaller than the social one, government intervention in the technological domain is aimed at bridging the resulting gap between private and socially optimal R&D efforts.

Governments may also want to stimulate private business or university R&D by strengthening the capacity of knowledge creation of universities in a given strategic field, helping the private business sector to increase its absorption capacity, or stimulating cooperation between industry and universities. If these policies are effective, public and private funding may be complementary, and increasing the former enhances the latter. However, there may be situations in which the government may have a negative impact on the innovation process. For example, government spending may crowd out private spending by increasing the demand of R&D and hence its price. Goolsbee (1998) and David and Hall (2000) argue that the major effect of government funding is to raise the wages of researchers. Higher research costs may shift business

R&D funding to alternative investments. As a result of this action, the real amount of R&D (adjusted for the higher cost of research) will be lower (even if the total amount of R&D is higher due to government funding). Guellec and Van Pottelsberghe (2003) mention several possible negatives of government R&D funding. One occurs when firms substitute public support for their own. In this case, there is no "additionality" coming from government funding. It is also possible that a firm starting a project thanks to government funding has the effect of deterring other firms from starting a similar project. This is a direct form of crowding out or displacement, which does not work through the price mechanism. Last, but not least, governments are less likely to allocate resources efficiently than market forces, which may generate distortions in the allocation of resources between fields of research. It may also distort competition between firms by supporting some at the expense of others.

Summarizing, the impact of knowledge and innovations on local productive systems is likely to be positive provided that the mechanisms for translating knowledge into technological change and into knowledge-intensive or high-tech jobs and regional growth are supported by the proper innovation systems. Regional innovation systems provide an economic, political, and social context, within which researchers, educators, companies, organizations, and governments can interact. Even if evidence for knowledge spillovers is found between institutional sectors involved in RIS, government policy may play an important role in reducing the market's failures and optimize the R&D effort.

1.6 Local Productive System and Innovations

1.6.1 Dynamic Externalities and Knowledge Spillovers

Another question to be raised is whether – and to what extent – the composition of industrial activity generates externalities that influence the region's knowledge creation and diffusion capacities. Innovation diffusion and its impact on regional economies depends on a number of factors such as the region's industry structure, firms' strategies and governmental policies, and linkages between regional actors and their roles in the generation and modification of innovations. This approach emphasizes the importance of a region's endogenous technological and innovative capabilities in generating knowledge externalities.

Glaeser et al. (1992) examine whether, and how, dynamic externalities are shaped by the composition of economic activity within a given particular geographical area. He considers three different types of dynamic externalities: MAR's externalities (Marshall 1890; Arrow 1962; Romer 1986), Porter's externalities (1990), and Jacobs' externalities (1969). All approaches suggest that innovation occurs more easily when geographical concentration and proximity are present.

MAR's externalities suggest that increased specialization in a particular industry within a specific geographic region facilitates knowledge spillovers across firms. Marshall (1890) observed that industries cluster geographically because intra-industry transmission of knowledge is favored by geographical proximity. Geographical proximity eases the interaction between individuals sharing similar specific competences and thus favors the diffusion of technologies, information, and knowledge. The main channel of diffusion is considered to be the local labor market (Gorynia and Jankowska 2007, p. 316). In the MAR models of externalities, innovators realize they cannot capture the entire return of their R&D efforts because some of their ideas are imitated or improved by other firms without compensation. The imperfect appropriability problem leads them to slow down R&D investment and justifies monopoly rather than local competition (Cincera 2003, pp. 10–12).

According to Porter's cluster-based theory of externalities (1990, 1998), the specialization of a local industrial structure, with many firms competing in the same industry or collaborating across related industries, tends to trigger innovation and learning processes. As Malmberg and Maskell (2002, p. 433) point out, "in such environment, chances are greater that an individual firm will get in touch with actors that have developed or been early adapters of new technology. The flow of industry-related information and knowledge is generally more abundant, to the advantage of all firms involved." According to Van der Berg et al. (2001), most definitions of the cluster "share the notion of clusters as localized networks of specialized organizations, whose production processes are closely linked through the exchange of goods, services and/or knowledge."[16]

The main difference between the MAR model and the model of Porter (1990) concerns the role of competition. Porter (1990) argues that although competition reduces returns to the innovator, it increases the pressure to innovate. This second effect is assumed to be far more important than the first one. Porter argues (1998) that competition pressurizes firms to innovate, and forces both local and foreign firms to develop radical innovations to survive and succeed in their market (Atuahene-Gima 2005; Zhou et al. 2005). Competition between local firms active within the same industry leads to rapid adoption and improvement of inventions and generates higher industry growth. MAR and Porter agreed that most important technological externalities occur within industry and that regional growth is mainly the result of regional specialization (Glaeser et al. 1992).

Contrary to MAR and Porter, Jacobs (1969) regarded inter-industry spillovers as the most important source of new knowledge creation. She argues that the

[16] Even though Porter's (1998, 2008) cluster approach has been vastly popularized at the regional and local level in the US, there are still problems with the definition of clusters. One of the simplest ways to define cluster is suggested by Doeringer and Terkla, according to whom industry clusters are "the geographical concentrations of industries that gain performance advantages through co-location" (1995, p. 225). Other authors, Barkley and Henry, define cluster as "a loose, geographically bounded collection of similar and/or related firms that together create competitive advantages for Memberfirms and the regional economy" (2002). Gibbs and Bernat (1997) further add to these definitions by identifying shared input needs and inter-relationships with suppliers and buyers. On the other hand, Swann and Prevezer put it very simply – they say that "geographical cluster is a collection of related companies located in a small geographical area. (…) Companies group together to take advantage of strong demand in the location, a large supply of scientific manpower and the network of complementary strengths in neighboring firms" (1998, p. 3).

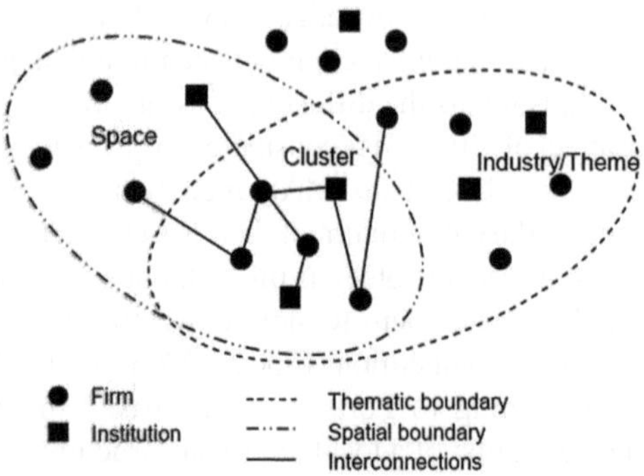

Fig. 1.3 Interconnections between firms and institutions in a cluster (Source: Menzel and Fornahl 2007)

agglomeration of different industries within an urban region fosters innovation due to the diversity of available local knowledge sources (Cincera 2003). Only in the context of industrial diversity rather than industrial specialization does the exchange of complementary knowledge lead to cross fertilization of ideas and new knowledge creation, which in turn may favor innovation and economic growth (Greunz 2004). In the debate between local monopoly and competition, Jacobs, like Porter, believed that local monopoly harms innovation.

The three main theories, which aim at explaining how dynamic externalities are shaped by the composition of economic activity within a particular geographic region, are illustrated in Fig. 1.3. The figure emphasizes the importance of flow of information between regional actors (firms and public institutions, i.e. government R&D and educational institutions) and the ability to adapt new innovations to firms' own environments. Furthermore, they suggest that the endogenous dynamic of innovation is highly dependent on the specificities of regions' technological capabilities (thematic boundary). Therefore, in order to understand the innovation process within the region, these are the characteristics that need to be considered. Technological boundary of a cluster "implies that only certain firms and institutions 'in a particular field' belong to the cluster." In other words, a certain "technological proximity" serves as the basis for various exchange processes and synergies. Thus, the cluster forms only a part of the regional production system. The geographic scope of a cluster (spatial boundary) can range from a single city or state to a country or even a group of neighboring countries. The spatial boundary delimits the firms of a cluster from firms located elsewhere. Figure 1.3 shows that the shape of the cluster depends on the interface between industrial and local dynamics (Menzel and Fornahl 2007; Albino et al. 1999).

The dynamic approach to clusters requires considering the changes in the spatial and technological boundaries of a cluster.[17] One of the factors and processes that influence the spatial and technological boundaries of a cluster is the ongoing

[17] Other factors that influence all firms and institutions in a cluster include legislation, regional cultures, social institutions and conventions, as well as regional assets. For example, a regional culture, like the one in the Italian industrial districts or in Silicon Valley, influences all firms to the same extent, whether they are part of a cluster or not.

process of globalization and increasing competition in knowledge-intensive sectors (Dicken 1998; Boyer and Durand 1997). Hence, important knowledge sources of the innovation of local firms stem from both local and global knowledge linkages. As a result, new firms and industries emerge, local institutions develop and change, and clusters change their boundaries.[18]

The idea of cluster dynamics was stimulated by the emergence of the evolutionary paradigm in economic geography (Boschma and Frenken 2006; Boschma and Martin 2010). One of cluster theory's founders, Michael Porter himself, did not have that much to say about cluster evolution, beyond a brief discussion of some of the reasons for the birth, growth, and decline of clusters in his book *On Competition* (1998, pp. 237–245).

This also explains the weakness of the common statistical approach to cluster analysis (based on either SIC or NAICS codes), in which regional or metropolitan-level cluster boundaries represent nothing more than empirical constructs. From the methodological point of view, such a situation leaves little difference between cluster concept and RIS (Porter 2000). However, the logic behind building RIS is to support and strengthen the localized learning process and promote "historical technological trajectories" of regions (Asheim and Coenen 2005). In this context, RIS is a necessary part of cluster development.

1.6.2 Dynamic Externalities and Industry Life Cycles

Given the fact that industrial clusters develop regionally along the industry life cycle, dynamic externalities generated by clusters will be affected differently by knowledge and skills. The aforementioned three types of dynamic externalities: the MAR, Porter's, and Jacobs', allow tracking the role of knowledge flows throughout industry's life cycle. The industry life-cycle model is based on a stylized description of the evolution of an industry and follows the logistics of an S curve, starting with the introduction of new products, followed by a period of strong expansion of production, which then levels off and eventually leads to a decline. Following the literature on the role of the agglomeration effects on knowledge spillovers, presented in Sect. 1.1.4, it is possible to assume that certain types of dynamic externalities assist the industry as it moves from a young to a more mature stage (Fig. 1.4).

Going into details, new industries – or industries at the introductory stage of their development – benefit mostly from diverse knowledge infrastructure and inter-industry knowledge spillovers. Therefore, Jacobs' externalities will be more important at this stage. The birth of a new industry typically follows radical innovations, which

[18] While cluster boundaries often fit political boundaries, they may also cross state and even national borders, especially in smaller States and nations and in cities located near borders (Porter 2008, pp. 204–205). Maskell (2001) defined the boundaries of the cluster as the "*fit* between the economic activities carried out by the related firms of the cluster on the one hand and the particular institutional endowment developed over time to assist these activities on the other."

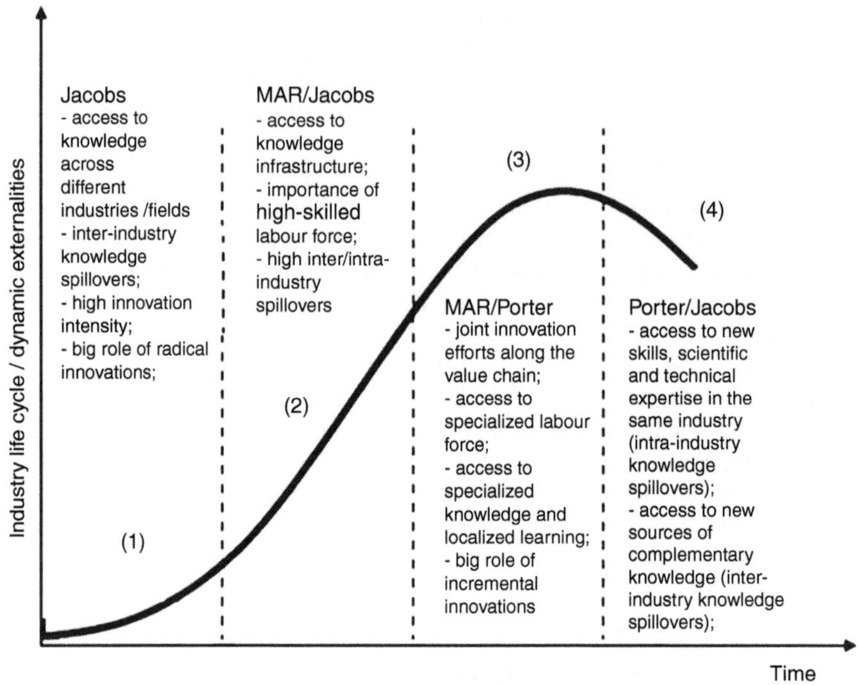

Fig. 1.4 Industry life cycles* and dynamic externalities (Source: Own elaboration *(from (1) introductory, (2) growth, (3) maturity and (4) decline phase of industry life cycle))

may originate outside of the particular industry or sector. Innovation intensity is high, as there are many unexplored technological opportunities (Neffke et al. 2009).

At the growth stage of industry development, production becomes more standardized, which opens up possibilities for firms to exploit their divisions of labor and economies of scale. At this stage of industry development, companies produce more or less similar products and get increasingly involved in price competition. This leads typically to a sharp drop in prices and a growth in production volumes. Both MAR's and Jacobs's externalities may be important at this stage.

At the stage of maturity, firms typically face vigorous price competition. Profit margins are reduced and technological opportunities get exhausted. In terms of innovation, longer jumps in technology are less likely and innovations are more of Arrow's nature (radical innovations are all but infeasible, as the industry has invested heavily in machinery and skill development that would become obsolete by dramatic discontinuities in technology). Major emphasis of R&D efforts is directed towards efficiency gains, which requires very specialized, industry-specific knowledge and skills (Gort and Klepper 1982). Such know-how is often of a strong tacit nature and is best acquired through processes of learning by doing and imitation. The focus on local possibilities to tailor education and training systems increases. The role of inter-industry spillovers becomes less significant, due to the industry's efforts on specialized knowledge and skills, which may penalize the high-diversity environment of a cluster. Both tendencies lead to a lowering of Jacobs' and an increase of MAR/Porter's externalities.

If the industry is unable to reinvent itself, it approaches a late state of maturity and will start declining if no radical changes are introduced, e.g. industries can rejuvenate after radical innovation with far-reaching consequences for the industry, which may take the industry back to a more infant stage. The latter requires concentrating on either upgrading the current knowledge sources or looking for new sources of innovations. Acquiring external sources of complementary knowledge is likely to have positive effects in terms of Jacobs' knowledge externalities. This depends, however, on the extent to which external knowledge can be efficiently absorbed and used. When the technological distance between different knowledge bases is too high, and these are divergent and unrelated, acquiring external knowledge is difficult. Consequently, new skills and scientific and technical expertise in the same industry can boost technological innovations and Porter's externalities.

Since the creation of MAR's, Jacobs's, and lately Porter's externalities, no consensus has been achieved in the literature concerning the role of externalities in explaining knowledge spillovers and innovations within industrial clusters. The empirical evidence of agglomeration economies remains so far largely inconclusive too. Apparently, industries at different phases of their life cycle need different externalities to generate innovations. This can be explained by many factors, such as features of the investigated industry, the historical period the study relates to, size of geographic region, and method of estimation (Rosenthal and Strange 2003). Therefore, it is important to carry out further tests for such dynamic externalities in case the selected industries have more specified environments.

1.7 Knowledge Spillovers, Technological Change, and Regional Growth

1.7.1 Knowledge Stock, Innovations, and Technological Catching Up

Technological change has been the engine of regional growth and prosperity for the past two centuries at least. The role of technological change for economic growth was tested empirically for the first time by the neoclassical theorists. Neoclassical economic models considered technological change as exogenous, falling as "manna from heaven" (Solow 1956; Swan 1956; Cass 1965; Koopmans 1965). Solow's model pictured technology as a continuous, ever-expanding set of knowledge that becomes evident over time (not as something that is specifically created by economic agents and their inter-linkages). The model assumes that technology and knowledge are public goods, so that "technological progress" can be accessed by anyone free of charge (Solow 1950). As a result, one should assume that all countries or regions must share the same pool of technology and converge in their growth. Naturally, if the technology is a public good, then whatever the source of differences in GDP per capita is, it is not technology.[19]

Similar growth patterns were developed by Gerschenkrohn (1962), Gomulka (1971), Abramowitz (1979, 1986), and Maddison (1982, 1991) in their "catch-up" concept. "Catch up" emphasizes differences in scope for imitation. Countries behind the world innovation frontier, they argue, could grow faster by copying technologies. According to Gomulka, "latecomers in industrialization should, and in fact do, tend to innovate faster than does the world's technology frontier, the latter defined as the region in which the world's best technology is employed." Therefore, "the falling of the relative technology gap between a less developed country and the world technology frontier is what is meant by technological catching-up" (1987, p. 379). He also emphasizes the role of human, institutional, and technological capabilities for favoring a more rapid transfer of foreign technology.

Following neoclassical growth theory, trade, FDI, and capital mobility may create a channel for diffusion on technology and catching up. For many catching-up countries, FDI plays an important role, as it is treated as an additional investment that increases domestic capital stock, know-how and technology (Runiewicz 2005; de Mello 1997). However, in order to assess the influence of FDI, technological catching up, and economic growth, one should look whether the externalities extended to domestic firms have any long-term impact on the host country. Studies have found that, despite the positive role of FDI, the tendency for poorer regions to grow at above-average rates in the longer period slowed down or even vanished. Herz and Vogel (2003) analyzed regional growth and convergence in a sample of 31 Central and Eastern European regions on the NUTS2 level over the period 1990–2002. They found that the regional disparity in Poland, Hungary, and the Czech Republic decreased in the first half of the 1990s, with a tendency of poorer regions to exhibit above-average growth rates and sample convergence. However, since then, the reduction of income and productivity dispersion came to a halt.

Furthermore, studies by Borensztein et al. (1998) and Kokko et al. (1996) found that the positive relation between FDI and productivity growth holds only when a host country or region has achieved a minimum threshold of human capital development. Other studies addressing this problem showed that local conditions, such as technological capabilities, human capital, and the development of domestic financial markets, are likely to play important roles (OECD 2001). Thus, the positive effects of FDI are not automatic but may be affected by various host industry and country characteristics. The fact that a country or a region is falling behind does not guarantee that technological catch up and growth will be achieved (Runiewicz 2005).

One of the most important contributions of Abramovitz was the emphasis put on the need for "social capabilities" to benefit from catch-up growth. In the spirit of

[19] Attempts to introduce it by presenting models with a separate technology-producing sector were made by Uzawa (1965), Phelps (1966), Shell (1967), and others. However, by the early 1970s, growth theory went out of fashion together with the idea of endogenous technological progress. As a consequence, these models had little, if any, impact on the empirical work that accompanied theoretical discussions of the 1960s.

Abramovitz (1986, p. 390),[20] regions that are technologically backward have the potential to generate high growth, provided their "social capabilities are sufficiently developed to permit successful exploitation of technologies already employed by the technological leaders." John Cornwall (1976) reached a similar conclusion. In the spirit of technological gap theories, he argued that technological differences between regions open up the possibility for countries or regions to catch up by imitating the most productive technologies ("advantage in backwardness"). He used GDP per capita to measure the gap in technology between frontier and the latecomer countries. He pointed out that the higher this gap is, the higher the potential for high growth will be in latecomer countries through successful imitation of superior technology developed elsewhere. In his view, to exploit this gap, countries needed to make additional investments and upgrade their stock of existing knowledge. In the same spirit, Gerschenkron (1962) suggests that the technological gap provides advantages for the successful imitation of advanced technology for latecomers, provided the latter fulfils the requirements.

This argument was further developed in catching up and regional convergence models provided by endogenous growth theory, which explains regional growth patterns with the production of knowledge or R&D (Romer 1990; Grossman and Helpman 1991; Aghion and Howitt 1992, 1998). Since technical progress is closely associated with knowledge emerging from R&D, endogenous technical change is considered to be generated by formal R&D activities. In fact, studies conducted by Chandler (1990) and Nelson and Winter (1982) prove that technological differences are in fact the prime cause of differences in GDP per capita across countries. Knowledge does not become a public good immediately. Innovations leak out only gradually. Firms will try to keep their innovations secret and protect them by patents. Thus, the capability for innovation and knowledge diffusion is assumed to be the major cause behind long-run differences in specialization patterns, competitiveness, and economic growth.

Other studies by Glaeser et al. (1995), Benhabib and Spiegel (2005), Eaton and Kortum (1996), Xu (2000), and Gottlieb and Fogarty (2003) identify the importance of endogenously determined technological change and level of education for catching up. Their studies show that education and R&D institutions are positive drivers of growth. As emphasized by Crafts (1996, p. 14), "a general implication of the new growth economics is that institutions and policy may have stronger effects on the growth rate than would have been predicted using the traditional neoclassical model" (1996, p. 41).

These new findings concerning technological catching-up theories do not mean that technological gap theories in the spirit of Gomulka (1971) and Abramowitz (1979, 1986, 1994) are not capable of explaining a region's catching up. Instead, they

[20] This is the way Abramovitz (1994) explained the failure of most industrialized countries to catch up with the US prior to World War II. Mainly, according to Abramovitz, a country that differs much from the technological leader in factor supply may sometimes find it difficult to absorb the leader country's technology.

suggest that technological change is the result of both investment in education and skills, capital stock accumulation, learning by doing, and R&D activity intensity. The region's capacity for technological catching up and long-term growth is determined both by its capacity to absorb and adopt externally generated knowledge spillovers and, secondly, by its capability to imitate (using its endogenous knowledge sources) and innovate (as it approaches the technological frontier).

1.7.2 Technological Change: Exogenously or Endogenously Determined?

As shown in the previous section, the process of technological catching up, in light of both neoclassical and endogenous growth theories, is the result of technological externalities, which in the case of the latter theory is a result of investments into knowledge and learning, whereas in the case of the former it is determined by technology transfer. Both theories emphasize the importance of capital accumulation; however, contrary to exogenous growth theorists, endogenous growth proponents identify its role with knowledge accumulation. Romer (1986) argues that, since knowledge is incorporated into a new capital, it is capital accumulation that improves the technology of each producer and the economy as a whole (via knowledge spillovers). On the other hand, Aghion and Howitt (1998) suggest technological knowledge itself is a kind of capital good (...). It can be accumulated through R&D efforts (p. 26). In fact, the role of R&D in technological caching up is perhaps the major difference between these two groups of theories.

The importance of the accumulation of knowledge and learning for technological catching up, emphasized by endogenous growth models, links this concept to the ideas of Arrow (1962) and Kaldor (1960). Arrow (1962) argues that "learning by doing" can generate spillovers, whether from tangible capital, human, or R&D expenditures. He advances the idea that knowledge creation and productivity gains (endogenously determined technological change) arise as a consequence of producing and investing. Productivity of a firm is an increasing function in cumulative aggregate investment in industry. Arrow's ideas draw on the observation that large-scale manufacturing in the technical field, as e.g. in aircraft manufacturing, has the potential to generate large experience and improvements in productivity over the production period. During the production process, workers try automatically to improve their efficiency. Thus, in the aircraft business, one can observe that after the design of a new model, the time required to build a new machine is reduced with each item produced, even though there is no obvious innovation introduced in the production process. Experience of how to produce the best goods and improvements in efficiency (thanks to accumulated knowledge and investments) occurred autonomously in the course of producing. Furthermore, learning by doing in a firm invokes increasing returns to the scale of a firm's production. The economy accumulates knowledge that increases along with the scale of production. Kaldor (1960) emphasizes that the growth of effective demand stimulates product and process innovations, and hence increases labor productivity and output growth. Furthermore, he argues that learning is closely related to firms' investment activities. The decline in physical capital would also cut down learning-by-doing activities and knowledge accumulation or technical change as well (Runiewicz-Wardyn 2008a).

Taking everything into account, technological change is determined both exogenously and endogenously. The faster accumulation of capital investments pushes up the relative demand for skilled labor and generates the process of learning (endogenously determined technological change). For example, machinery and tools at a high technical level cannot be used at their full potential without sufficiently trained personnel. On the other hand, a well-trained labor force cannot be properly used if its skills and creativity do not match the corresponding material capital. Therefore, capital accumulation enlarges the stock of knowledge and, as a result, enhances total factor productivity growth. Capital-driven growth is an important stage in the transition towards, what Porter (1990) refers to as innovation-driven competitive advantage, which, however, does not happen automatically. Yet, the transition to the innovation-driven stage of growth occurs when a country or region is human-capital-abundant and has active R&D. The lack of "critical mass" of R&D funding and the labor force are among the main difficulties of the regions in achieving sustainable and innovation-driven growth.

1.7.3 Spatial Knowledge Spillovers and Endogenous Growth of Regions

Knowledge spillover is the second key element to explain increasing returns to scale and consequently the endogenous growth model. Knowledge spillovers are central in Romer (1986, 1990), Lucas (1988), and Aghion and Howitt (1992), and their scope is critical for the long-run distribution of incomes in the models of Grossman and Helpman (1991), Lucas (1993), and Howitt (2000). The increasing role of geographical proximity in shaping economic interactions (labor mobility, inter-firm linkages, etc.), knowledge spillovers, and innovative propensity triggered the "new" economic geography literature. The "new economic geography" models conceived growth as endogenous, generated by the advantages of the spatial concentration of activities and agglomeration economies. After all, geographical proximity matters in transmitting knowledge, because, as Glaeser et al. (1992, p. 1126) observe, intellectual breakthroughs must cross hallways and streets more easily than oceans and continents (Audretsch and Feldman 2004, p. 2719). This suggests that knowledge and innovative activities are strongly polarized in geographical space. However, the geographical concentration of knowledge cannot be narrowed to regional boundaries, as Krugman (1991) expresses, "[...] knowledge flows are invisible; they leave no paper trail by which they may be measured and tracked" (1991, p. 54). This shows that any attempts to measure the impact of knowledge spillovers on technological change and regional growth is extremely difficult, if not impossible.

Knowledge spillovers arise to the extent that the knowledge created by one regional agent becomes a public good (e.g. if it is not patented). The external benefits from the creation of technological knowledge that flow to parties other than the inventor may have an impact on income convergence across regions. Strong technological spillovers favor convergence, while weak spillovers can lead to divergence if the rate of technological change varies across regions. The capacity of regions to absorb, diffuse and exploit external knowledge depends on the level of development of their knowledge infrastructure and social capacities in particular (as discussed in previous sections of this book).

Empirically, the literature of the new growth theory considers the measures of the impact of geographically bounded knowledge spillovers on technological change and regional growth. To explain the increased divergence in the distribution of economic activity between countries and regions, Romer (1986) relied on models based on increasing returns to scale in production within a spatially distinguishable unit. In fact, it was assumed that the externalities across firms, and even industries, would generate increasing returns in production (dynamic externalities). To the extent that these externalities have geographic limits, they have implications for regional long-run industrial growth, e.g. through regional specialization/employment, level of productivity, and income per capita growth. The presence of spillover effects and related externalities cannot be limited to a given regional economy (intra-regional knowledge spillovers). Regional capability to innovate and prosper may be related to inter- or extra-regional knowledge spillovers. That, on the other hand, is determined by institutional spillovers between home and neighboring areas, as well as by the shared set of historical and cultural factors. Just as we do not know the exact cause of knowledge spillovers, Audretsch and Feldman state, "How and why does knowledge spill over is more than an academic question" (2004, p. 2735). The exact scope and size of the impact of knowledge spillovers on regional growth requires consideration of broader territorial inter-linkages and potential synergies on innovation and regional convergence. Similarly, the effects of technological spillovers must extend beyond factories and research centers, and consider the broader scale of social and economic aspects.

1.8 Summary and Conclusions

The influence of knowledge on technological change is best observed through the process of invention, innovation, and diffusion of knowledge. The process of innovation is determined by the context of a particular industry and its specific knowledge base. The impact of innovation on technological change depends on the one hand, on the rate of diffusion of new technological knowledge across economic actors and institutions involved and on their ability to absorb this new knowledge, on the other. The latter depends critically on the stock of already existing knowledge and previous innovations within these organizations. Furthermore, the process of knowledge creation and diffusion depends on the efficiency of the territorial organization of knowledge production. Knowledge spreads more rapidly in agglomerated urban areas and in close proximity to major universities. Interactive, huge, and diverse social capital makes large agglomeration regions. However, the importance of spatial proximity for knowledge spillovers and diffusion is further determined by the specific industrial context and type of knowledge involved in the process of innovation.

There aren't any direct measures of the rate of technological change; therefore, its further measures will be linked to several proxies, such as input and output-based measures. The first group of measures relates to R&D intensity and patenting activity, whereas the second group refers to productivity and employment growth variables. Regions that bring together both high R&D intensities and number of patents achieve higher rates of productivity and increase their employment in the high-tech and knowledge-intensive sectors. However, the mechanism for translating new knowledge into higher productivity and knowledge-based growth is not automatic. It is likely to happen if it is supported by the proper technological infrastructure or "innovative system," which enables and enhances the process of lifelong learning and the diffusion of new technological knowledge throughout local industrial structures (clusters). Regional industrial clusters and innovation systems are closely linked and mutually reinforcing. Regional technological infrastructure supports local industrial clusters with collaborative public research institutes, high-quality academic institutions, and vocational training organizations.

New economic growth theories and "new economic geography" indicate a high potential for explaining regional growth patterns. Many empirical findings have proved that geographically bounded knowledge spillovers and technological differences are, in fact, the prime cause for differences in GDP per capita across countries and regions (Glaeser et al. 1995; Benhabib and Spiegel 2005; Eaton and Kortum 1996; Xu 2000; and Gottlieb and Fogarty 2003). Yet, their results should be put in the specific local community context. For regions with low technological capabilities, technology transfer through the faster accumulation of capital investments (exogenously determined technological change) is crucial in order to induce endogenously determined technological change, e.g. by the process of learning by doing.

Therefore, further attempts are still required to produce a theoretical approach that would incorporate relevant territorial features of knowledge spillovers (the efficiency of RIS) and technological externalities into macroeconomic regional growth models. Development of such a framework would greatly contribute to the further development of a new growth theory in space.

Chapter 2
Innovative, Technological, and Growth Capacities of the EU Regions

2.1 Introduction

The role of regions as engines of economic development and growth has been widely recognized in recent years, and abundant documentation now exists of many of the successful economic examples of this phenomenon in different parts of the world. Critical elements of the economic success of regions depends on the capabilities of the local level to upgrade its productive structures and to generate, diffuse, and apply knowledge in the production of highly innovative and knowledge-intensive products and services. Accelerating technological change and moving up the technological ladder, from low to high value-added industries is a prerequisite for sustaining economic competitiveness. These industries create good jobs, expand production and trade, and drive continuous innovation. It is therefore not surprising that most developed economies increase their share of knowledge and research-intensive industries such as biotechnology, ICT, advanced manufacturing, and advanced business services.

The following chapter analyzes how EU regions perform in terms of their capacity to create, transfer, and diffuse new knowledge. The first section provides some useful facts and figures that highlight the heterogeneity of the European regional landscape regarding wealth and knowledge-creation capacities such as GDP per capita growth. Drawing on the regional Eurostat database and other economic data, the second section of the chapter looks at the economic structure and dynamics of the EU regions, starting from productivity growth and extending it to the analysis of technology and knowledge-intensive employment. The third section analyzes the factual data on innovation and knowledge absorption, diffusion, and creation capacities of the EU regions. The fourth section of the chapter benchmarks the EU regions according to their growth and innovative capacities. Finally, the last section discusses the spatial distribution of technological and innovative potential of the EU. The chapter ends with the summary and important conclusions.

M. Runiewicz-Wardyn, *Knowledge Flows, Technological Change and Regional Growth in the European Union*, Contributions to Economics,
DOI 10.1007/978-3-319-00342-9_2, © Springer International Publishing Switzerland 2013

2.2 General Economic Conditions in the EU Regions

2.2.1 GDP Performance in the EU Regions

GDP per capita is the most frequently used indicator to evaluate and compare the economic performance of regions in terms of wealth creation. This indicator, expressed in purchasing power standards (PPS), has also been used to assess the heterogeneity of the European regional landscape and the average economic situation in each of the EU Member States regions.

Map 2.1 clearly illustrates an unequal distribution of wealth creation across the EU. Firstly, it reflects a high concentration of wealth creation (above the European average GDP per capita for 2009) only in a limited number of regions, extending from the North EU regions to the Benelux, western Germany, western Austria, and ending in the northern part of Italy. Secondly, the three regions with the highest GDP per capita over the past years have been Inner London, Brussels, and Luxembourg, followed by Hamburg, Île de France, Wien, Uusimaa, Stockholm, and Madrid. Inner London and Bruxelles are the wealthiest regions of Europe with a GDP per capita of more than twice that of the European average. Thirdly, most regions belonging to the new Member States as well as the southern European periphery, such as the Portuguese, Spanish, southern Italian, and Greek regions, are characterized by relatively low levels of GDP per capita.

Furthermore, there are significant differences in the capacity to create wealth within the EU Member States of EU national boundaries. The capacity of wealth creation of the capital and highly agglomerated regions is relatively better in Hamburg, Ile de France, Wien, London, and Stockholm in comparison with Mecklenburg-Vorpommen, Corse, Brugenland, and the North East and Mellensverige regions. Similarly, the capacity of wealth creation of the capital regions is relatively better, e.g. Budapest, Prague, Warsaw, and the Bratislava regions in comparison to Észak-Alföld, Strední Morava, Lubelskie, and Východné Slovensko. As pointed out in the theoretical part of this book (Chap. 1), the urban concentrations of capital cities lead to economic growth through local knowledge spillovers, which in turn affect local accumulation of capital and agglomeration economies, and, as a result, further growth. Consequently, this has made rich regions become richer and poor regions poorer.

From the dynamic perspective, Map 2.2 indicates that all regions performed well. Map 2.2 shows the extent to which per capita GDP changed between 2000 and 2009 on average in the EU regions. The map shows that economic dynamism is well above average in the southwestern, eastern, and northern peripheral areas of the EU, particularly in the EU-10 Member States. Based on the most recent estimates released by Eurostat (April 13, 2011) the gap between the richest and poorest EU regions has narrowed since 2000. In 2009, only 65 regions had a GDP per capita below 75 % of the EU average, in comparison to 69 regions in 2000. This represents that 119 million people stayed below 75 % of the EU average GDP per

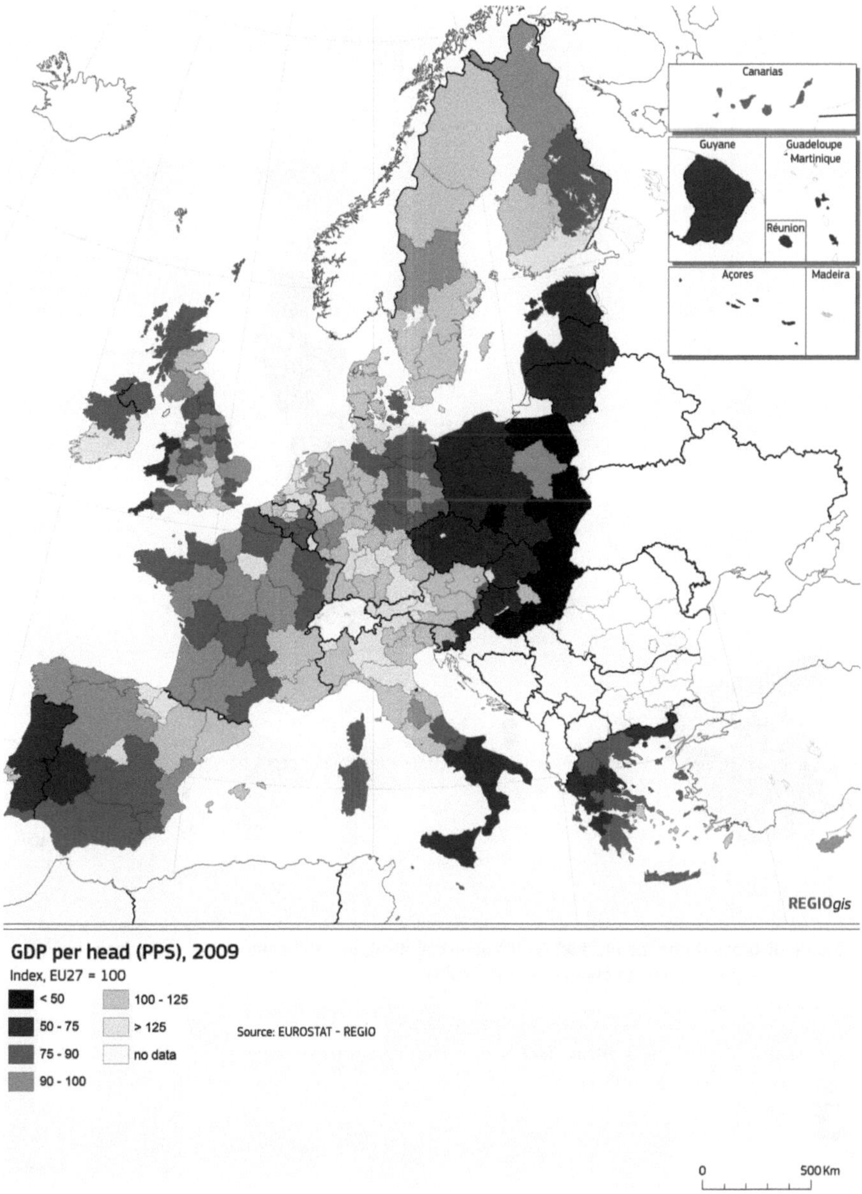

Map 2.1 GDP per head (PPS) by NUTS2/3, 2009 (Source: Eurostat – REGIO)

capita, compared with 131 million people in 2000. As a result, the gap between the richest and poorest EU regions has narrowed since 2000 (see Maps 2.1 and 2.2).

Among the EU-15 Member States, strong growth was particularly evident in Spain, parts of The Netherlands and Greece, as well as the north of Finland and

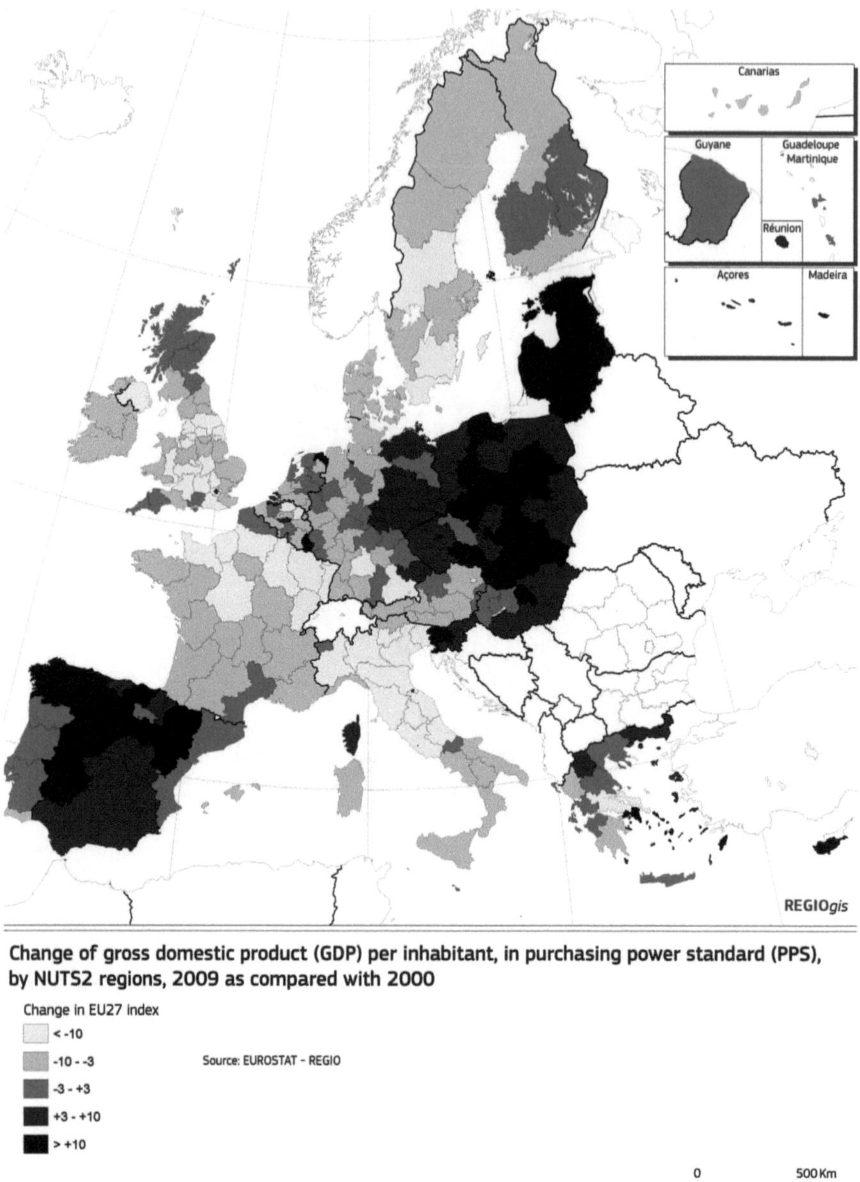

Map 2.2 Change of GDP per capita (PPS) by NUTS2/3 regions 2009 compared to 2000 (Source: Eurostat – REGIO)

Sweden. On the other hand, weak growth that started several years ago is persisting in several EU-15 regions belonging to Italy, France and Portugal, Germany, Sweden, and the UK. Among the EU-10 Member States regions of the Baltic

States, regions of Slovakia, the Czech Republic, and most of Poland have seen growth markedly above the average.

Eurostat-based regional data reveals that the catch-up process of EU-10 countries with the EU average was of the order of 1.7 % age points per year between 2000 and 2009. This fast process of catching up was driven by economic integration and restructuring of national economies. GDP per capita in the EU-10 Member States rose from 50 % of the EU-25 average in 2000 to over 60 % in 2009. In 2008, performance was particularly strong, above 3 % points. It is also important to mention that the fast catching up in the second half of the period under analysis can be explained partly by the fact that the economic and financial crisis struck first in the EU-15 Member States, some of which, like Ireland, Italy, and Denmark, were already in recession in 2008. On the other hand, among the EU-10, only Estonia and Latvia already had negative volume growth rates in 2008, and the full effects of the crisis became apparent only in 2009. EU average of GDP per capita (in PPS) dropped by 6 % between 2008 and 2009 (*Eurostat. Statistics in Focus* 41/2012). Regional GDP per capita dropped sharply in 2009 compared with 2008 in all EU Member States except for Poland (11 out of 16 Polish regions achieved absolute increases in 2008–2009).

The crisis affected mostly manufacturing, construction, and exports (including tourism) activities. Regions with the lowest per capita GDP and capital city regions suffered smaller setbacks than the EU as a whole, resulting in even bigger regional inequalities at the intra-national level in most EU Member States.

However, this fast catch up in terms of GDP per capita resulted in a sharp increase of regional disparities in EU countries. Economic liberalization and integration have favored all of the above regions with significant agglomeration economies, vast concentrations of skilled labor, and vigorous demand.

This first positioning of some European regions in terms of their capacities to create wealth and catching up naturally leads us to the following question. What determines the ability of EU regions to catch up? To what extent does productivity growth (which in turn depends on technological advances, which further depends on the quality of human capital and engaging in R&D and knowledge-intensive activities) explain a region's capacity to prosper?

2.2.2 *Regional Productivity and Employment in Knowledge-Intensive Sectors*

The contribution of any employee in an industrial sector to the regional economy is best measured by the gross value that the individual adds (GVA). As a "residual," the GVA indicator measures the sum of incomes generated by the process of production and thus can be used for calculating technological efficiency. With reference to the previously discussed GDP trends, GVA is a method of measuring the productivity of a given economy. Under the European System of Accounts 1995, the term GDP is equal to GVA plus taxes on products less subsidies on products, i.e. at market prices. At the aggregate level, data on real GDP growth has been available in the Eurostat database only since 2000. Map 2.3 shows the average dynamics of GVA in the EU regions in the period 2000–2008. The average rates of growth observed in the EU-10 regions were significantly higher than those of their EU-15 counterparts. This should lead to a sizeable decrease in the productivity gap between these two groups of regional economies. In fact, the Baltic States (Latvia, Estonia, and Lithuania) have observed above 7 % of average growth in GVA, whereas the regions of the Czech Republic (Strední Cechy), Poland (Mazowieckie, Lubuskie, Podkarpackie, Śląskie, Łódzkie, Małopolskie, Pomorskie, Kujawsko-Pomorskie, Wielkopolskie), Slovakia (Západné Slovensko, Bratislavský kraj), and Hungary (Közép-Magyarország) have experienced above 5 % growth of GVA during the above-mentioned period. This might suggest that the processes of economic integration and liberalization in these EU-10 regions contributed to the highest efficiencies in terms of organizational improvements and acceleration of technology transfer from foreign investors. The rest of the EU-10 regions have recorded above 3 % average growth of GVA.

On average, productivity levels in capital cities and in large and dense agglomerations were much higher relative to the countries' national averages. This concerns the Polish Mazowieckie and Śląskie regions, the Slovakian Bratislavský and Západné Slovensko regions, Czech's Praha and Strední Cechy regions, and the Hungarian Közép-Magyarország and Közép-Dunántúl regions. High productivity in these regions relative to their national averages can be explained by the share of the manufacturing sector in their total employment, their market structure, the extent of competition, the level of communication infrastructure, and access to education and training.

Despite the relatively rapid catching-up process observed in recent years, labor productivity levels in the new EU Member States are still well below those observed in the EU-15 countries. The largest difference can be seen in manufacturing, whereas construction and market services seem to be lagging behind less. On average, the total level of productivity in industry is 2–3 times lower in the EU-10 than in the EU-15. Smaller countries such as Cyprus, Malta, and Estonia, along with the capital regions in the Czech Republic, Hungary, and the Republic of Poland have the highest productivity levels (Eurostat 2011).

All the EU-15 regions were found to have productivity above the EU's average productivity. The most productive regions include Groningen in the north of The Netherlands as well as two other Dutch regions, Zeeland and Zuid-Holland, southern and eastern Ireland, Brabant Wallon, Antwerpen and the capital region in Belgium, Sterea Ellada in Greece, the Övre Norrland in the north of Sweden, and the regions of Stockholm and Hamburg. Portuguese industry has half the EU's average productivity – the lowest among the EU-15 group, followed by the Greece, Spain, and southern Italy.

One of the reasons why an average worker in one of the EU-10 Member States still only produces about half the amount of goods and services that are produced in

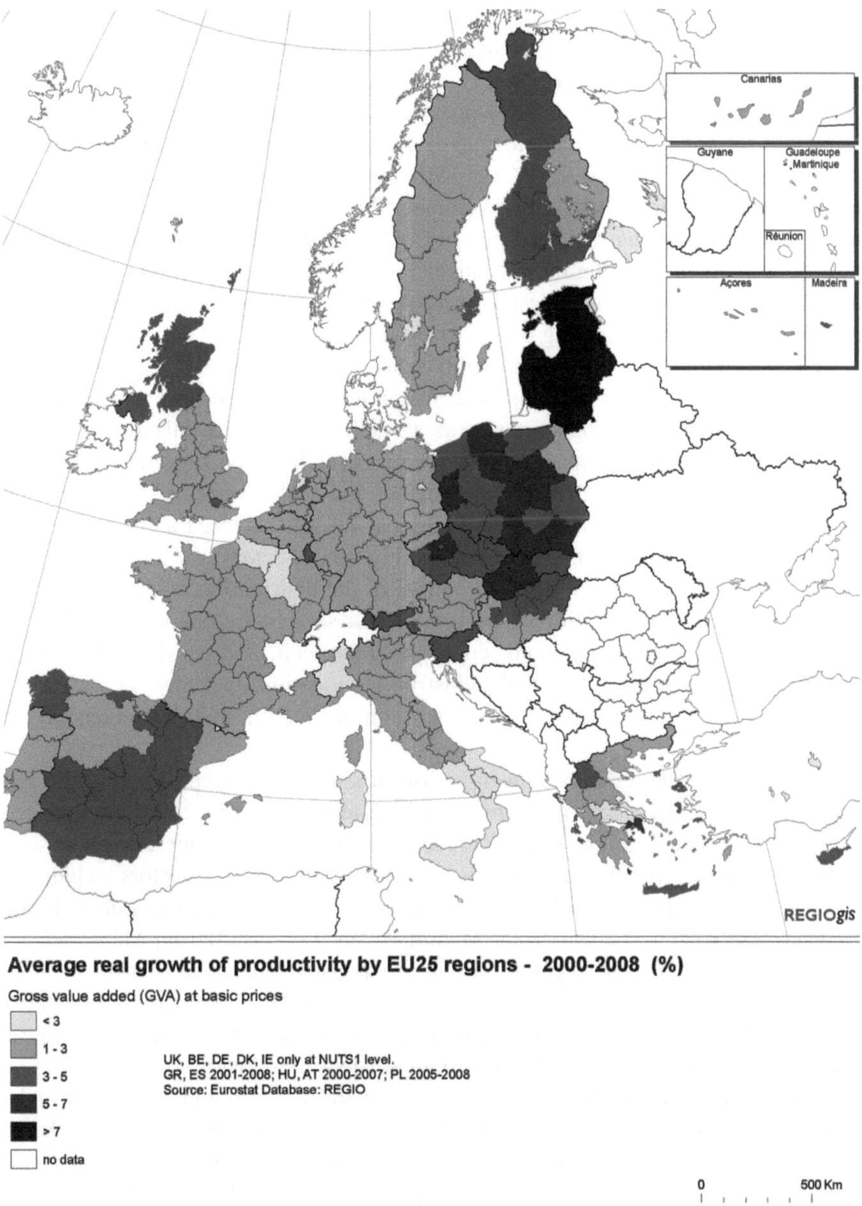

Map 2.3 Average real growth of productivity by EU-25 regions – 2000–2008 (%) (Source: Eurostat – REGIO)

the pre-enlargement EU is the difference in capital intensity (Kolasa 2005). Moreover, according to endogenous growth, productivity level is a function of the stock of accumulated knowledge. Since new Member States are surely behind a technology frontier, the positive externalities in their regional production function are expected to emerge via knowledge spillovers and transfer of technologies from the more developed EU-15 regions.

In a nutshell, while GVA acts as a proxy for technological efficiency, the latter is important for growth and technological catch up of regions, but there is substantial variance across EU regions in GVA growth. Furthermore, there is a large heterogeneity across countries in what this "residual" TFP component entails. When talking about technological change as a driver of GVA growth, one has to carry out a more precise assessment of what these changes are really about. The next sections will analyze in more detail how the catching-up regions are doing on various knowledge economy dimensions affecting productivity growth.

A key condition for fast productivity convergence of EU-10 regions towards more productive EU-15 regions was manufacturing investment intensity and capital-embodied technology transfer as a result of trade liberalization and FDI inflows. However, their impact on technological catch up has been determined by the ability of the regions to absorb and diffuse new technologies as well as by the extent to which foreign investors and national reforms favored a structural shift towards more knowledge-intensive sectors (high-tech sectors).

Map 2.4 provides a more detailed outline of the distribution of employment in high-tech sectors across European regions. These sectors are defined according to their high R&D intensity, and comprise high-tech and medium high-tech manufacturing (see the full list of industry classification in Annex 1) as well as high-tech knowledge-intensive services, such as R&D, and computer-related activities.[1]

As a general rule, employment in high-tech sectors is dispersed across the EU regions. Average share of employment in high-tech sectors in 2000–2008 ranged from 0.9 % in central Greece to 10 % in the capital region of Sweden. As can be seen in Map 2.4, the regions with the highest employment in high-tech sectors include the national capital regions of EU Member States (Île de France, Etelä-Suomi Praha Bratislavský, Közép-Magyarország, Madrid, Wien, London, Berlin, Mazowieckie, and Bruxelles). All these regions on average registered a value of over 5 % of employment in high-tech sectors between 1999 and 2007. These densely populated capital or city districts have higher R&D and patenting intensity and better absorptive capacity in order to be able to benefit from knowledge spillovers.

Beyond this concentration in capital cities, there was also a high share of high-tech employment in large metropolitan regions in Germany (Baden-Württemberg, Hessen, Bayern, Hamburg), Sweden (where the major regions were Östra Mellansverige, Sydsverige, and Västsverige), Finland (Pohjois-Suomi), the UK (East of England), France (Rhône-Alpes and Midi-Pyrénées), Italy (Lombardia

[1] Employed persons are persons aged 15 and over in high-tech and knowledge-intensive services sectors (high-tech KIS-sector). The knowledge intensity reflects the integration with a generic or service-specific science and technology base. It can be seen as a combination of knowledge embedded in new equipment, personnel, and R&D intensity, Eurostat REGIO.

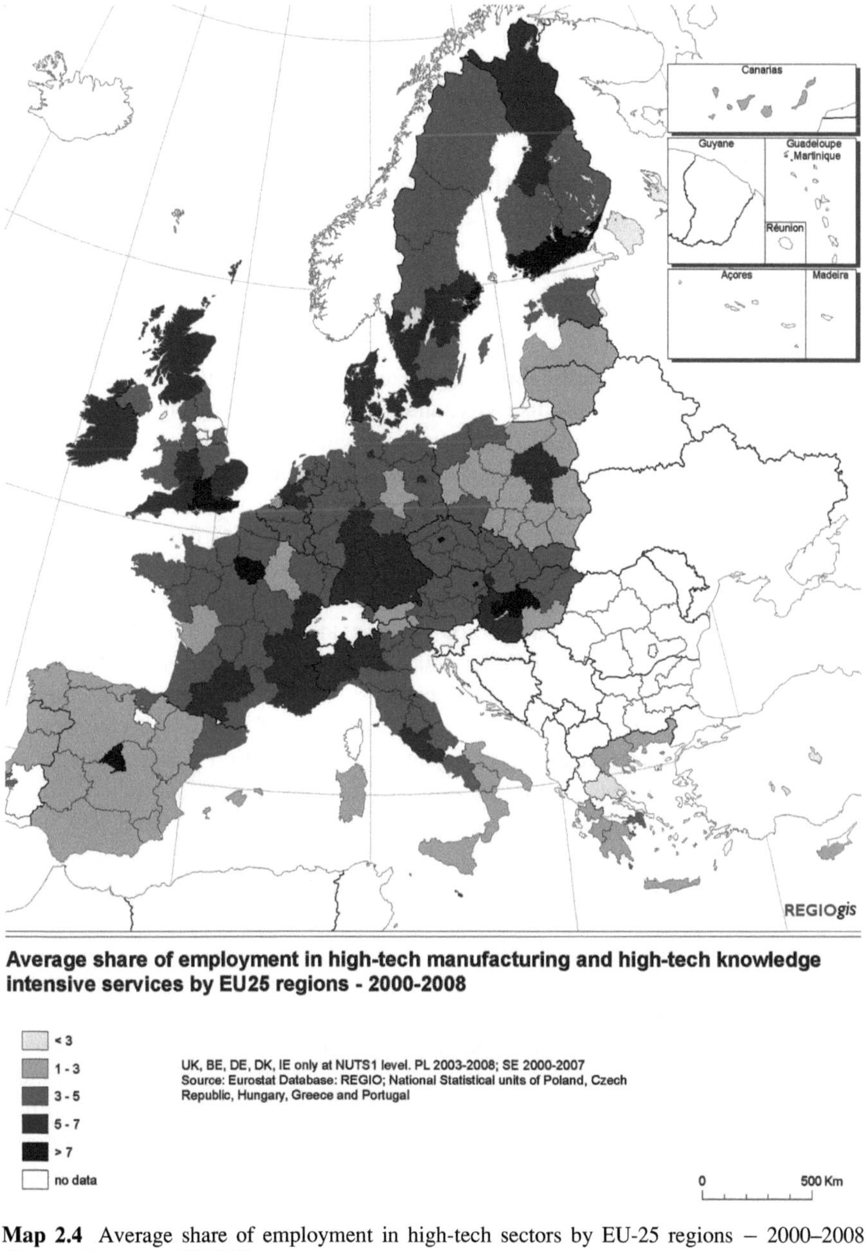

Map 2.4 Average share of employment in high-tech sectors by EU-25 regions – 2000–2008 (Source: Eurostat – REGIO)

and Piemonte), and Hungary (Közép-Dunántúl and Nyugat-Dunántúl). Finally, there is a cluster of high-tech regions that stretches from Luxembourg, through south-eastern Belgium up to East Flanders (Flemish Brabant) in the north of

Belgium, with two regions in the southern and western part of The Netherlands (Noord-Brabant and Zuid-Holland).

2.3 The Innovative Potential of the EU Regions and the Efficiency of RIS

2.3.1 Innovative Potential Indicators

The previous section has highlighted significant differences regarding the average growth of GDP per capita, productivity, and high-tech employment within the EU and its regions. The following section in turn overviews some aspects regarding the structure and spatial concentration of innovative potential of the EU. Two kinds of innovation potential capacities are investigated in the present section: patent data and R&D efforts. Patent applications to the European Patent Office (EPO) act as a proxy for the knowledge creation capacity.

Globally, German, French, and the UK regions altogether accounted for half of the average number of EU patent applications to the EPO during the period 2005–2008. The southern EU member states regions contribute with some 12 %, whereas EU-10 Member States regions comprise only 3 % of the total number of patent applications. The weak innovative performance of the EU-10 countries is mainly due to the fact that at the beginning of the transition period, innovative activity was almost absent, due to the obsolete technological infrastructure. Their national innovation systems were undergoing major restructuring, and RIS were just emerging.

In terms of geographical distribution of patent applications, there are significant variations within the regional structure of the leading countries regarding patents. Regions that are active in patenting are often situated close together, forming economic clusters. Their high performance and the concentration of innovative potential may be attributed to both local externalities and inter-regional knowledge spillovers. This is the case, for example, in the southern part of Germany (Rheinland-Pfalz, Baden-Württemberg, Bayern, Hessen), the south-east of France (Île de France and Rhône-Alpes), a northern province of The Netherlands (Noord-Brabant), the western region of Austria (Vorarlberg), the southern and western regions of Finland (Etelä-Suomi and Länsi-Suomi), and the southern and western regions of Sweden (Stockholm, Sydsverige, Västsverige) (Map 2.5). All these regions recorded on average more than 200 applications for every million inhabitants during the period 1999–2007.

The best performing regions among the EU-15 southern countries are the Italian regions of Emilia-Romagna, Lombardia Piemonte Veneto Friuli-Venezia Giulia, which submitted on average more than 100 applications during 1999–2007. The EU-10 regions submitted on average less than 25 patent applications during the

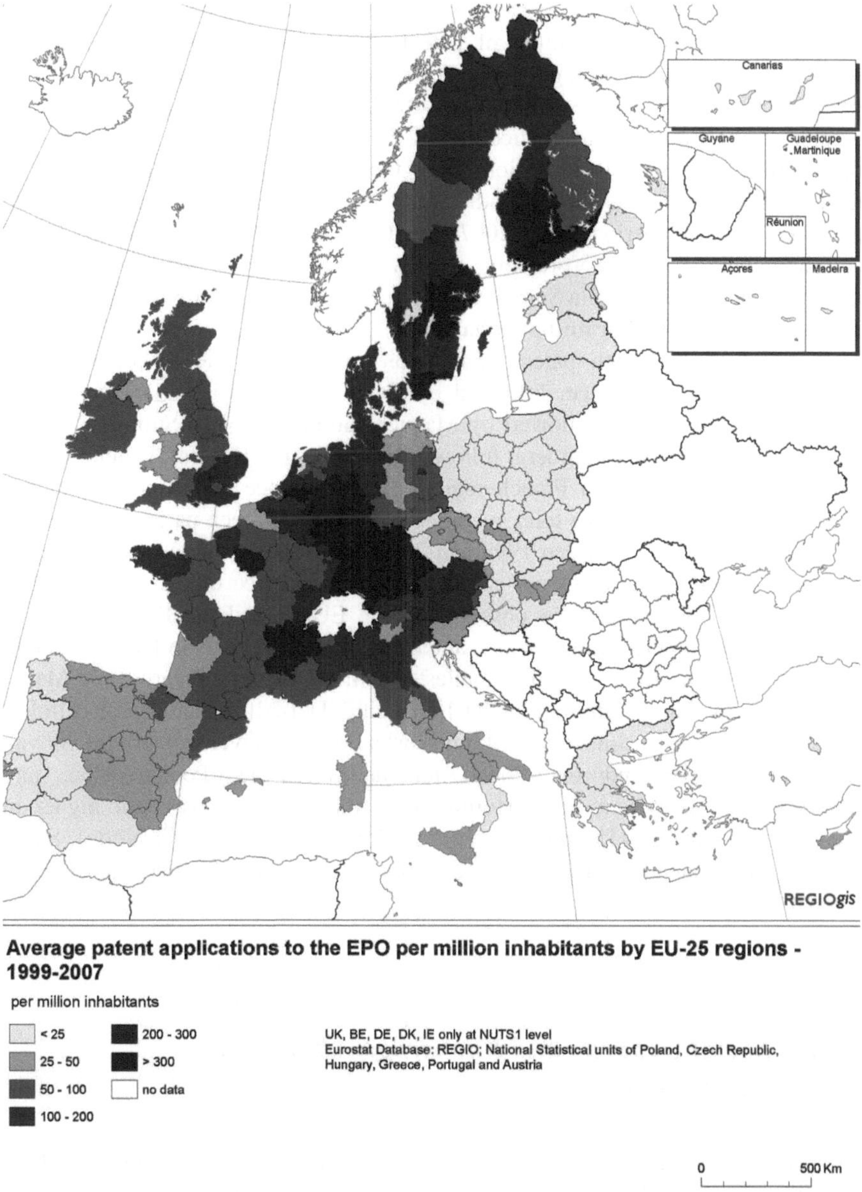

Map 2.5 Average patent applications to the EPO per million inhabitants by EU-25 regions – 1999–2007 (Source: Eurostat – REGIO)

analyzed period (except for Slovenia and Hungary's capital region Közép-Magyarország).

Some similar trends must be highlighted regarding the spatial concentration of innovation potential across the EU regions towards their relative national average values. Figure 1 in Annex 2 presents the best and worst regions within each EU country in terms of the number of patent applications per million inhabitants relative to the national average value. The ranking is based on average number of patent applications over the period 1996–2007. It clearly illustrates that there is a greater dispersion of patent applications at the regional level than at the national one. Furthermore, the EU-15 regions show the highest regional dispersion within their national economies in this indicator with respect to the EU-10 group. On average, over the period 1996–2007, the German Baden–Wattemburg, Dutch Noord-Brabare, Austrian Voralberg, the Swedish capital Stockholm, French Ile de France, and Finnish Etela Suomi regions occupied the highest positions in terms of patent applications per million inhabitants.

In terms of high-tech patent applications per capita, the Swedish regions (Stockholm and Sydsverige, followed by Östra Mellansverige Övre Norrland), Finnish (Pohjois-Suomi, followed by Manner-Suomi and Länsi-Suomi), German (Bayern and Baden-Württemberg, followed by Berlin and Hesses), Dutch (Noord-Brabant and Utrecht), and French regions (Ile de France and Bretagne Rhône-Alpes) are the leaders in the total average number of high-tech patent applications submitted to the EPO between 1999 and 2007. In all these regions the number of high-tech patent applications per capita is more than twice that of the European average. The opposite can be observed in the EU-10 and the EU-15 southern regions. The best performing regions among the EU-10 regions in terms of patent applications to the EPO were the smallest countries: Estonia, Cyprus, and Malta, followed by the Hungarian regions (Dél-Dunántúl, Dél-Alföld, and Észak-Alföld), Czech regions (Praha, Jihozápad, Moravskoslezsko, and Severozápad), the Slovak Bratislava region, and the Polish regions (Lubuskie, Podkarpackie, and Mazowieckie). However, the performance of these regions was still higher in comparison to the EU-15 southern regions: Portugal (Algarve and Norte), Spain (Galicia, lles Balears, Extremadura, Canarias, Castilla-La Mancha), and Greece (Sterea, Ellada, Thessalia).

2.3.2 Social Capacity and Knowledge-Absorption Determinants

As previously stated, knowledge creation and innovation may be the outcome of the region's own research and innovative efforts, or be accessed from external sources such as firms or R&D institutions located in other regions. It is therefore essential for these regions to develop their knowledge absorption capacities or, as Abramovitz put it, "social capability" in order to facilitate innovation and the implementation of technology spillovers. In other words, knowledge spillovers occur if regions have the capacity to integrate them.

The share of human resources involved in S&T of the working age population and the number of students in tertiary education are proxies of the EU region's endowment of "social capability" (Table 2.1).

Table 2.1 gives an overview of the current absorption capacities of the EU regions. Absorption capacity is particularly strong in the north of Europe. Without exception, all Swedish, Finnish, and UK regions, and also German regions, Denmark, along with two capital regions of Poland and the Czech Republic combine a high share of the S&T population and have been noted to have levels of educational attainment that are above the European average

In general, the EU-15 regions have better developed absorption capacities than the EU-10 states regions, with the exception of the capital regions of Slovakia (Bratislavsky kraj), Poland (Mazowieckie region), the Czech Republic (Praha region), and Estonia. The Swedish, Finnish, and Dutch regions, followed by a few Spanish, French, German, and Austrian capital regions (Madrid, Ile de France, Berlin, and Wien) have the highest share of S&T human capital and level of tertiary education ratio. Moreover, compared to the French case, all Austrian regions have higher participation rates in tertiary education in comparison with S&T schooling. The German and Belgian regions are in a high position both in terms of participation in tertiary education and S&T human capital.

The majority, however, of French, Austrian, Spanish, and a few Italian regions has only moderate absorption capacities. The absorption capacity of most South Mediterranean regions (Italian, Greek, and Portuguese regions) is generally low.

A common feature of all Polish regions is their low S&T human capital and the high number of tertiary education students. A similar situation is applicable to several other EU-10 regions, e.g. Hungarian and Slovakian regions, along with Latvia and Lithuania, for which the general weakness is low S&T capital.

Following the classification of the KIT (*Knowledge, Innovation, Territory*) project, conducted within the framework of the ESPON 2013 Programme, Map 2.6 presents 'Scientific regions,' defined as those regions which simultaneously show higher than average research activity and higher than average quality of human capital. Four indicators capture the level of research activities: R&D expenditures per capita; percentage of employees in R&D; number of patents per capita for all economic sectors; number of patents per capita for the subsample of high-tech sectors.

The composite indicator is calculated as the unweighted average of the re-scaled scores for all indicators within the respective dimension (*KIT Final Report 2013*). The "scientific regions" are classified as those showing values greater than zero for both indicators. Regions showing values greater than zero for the human capital indicator, but less than zero for research activity are labelled 'human capital-intensive regions.' On the contrary, regions characterized by values greater than zero for research activity and less than zero for the human capital indicator are indicated as 'research-intensive regions.' Finally, regions showing values less than zero for both indicators are defined as regions with no specializations in knowledge activities. The KIT project identified 74 scientific regions, 30 research-intensive regions and 52 human capital-intensive regions. However, the biggest number of

Table 2.1 "Social capability" and knowledge absorption capacities of the EU regions (2008)

y	X	Students in tertiary education by NUTS2 (2008)					
		High (S > 111877.67)		Medium (41513.67 < S < 111877.67)		Low (S < 41513.67)	
Human resources as S&T by NUTS2 (2008) as % of totally economic active population	High (H > 30.13)	Vlaams Gewest Région wallonne Praha Denmark Baden-Württemberg Bayern Berlin Hessen Sachsen Ireland Com. Madrid Île de France	Rhône-Alpes Közép-Magyarország Zuid-Holland Wien Mazowieckie Manner-Suomi Etelä-Suomi East of England London South East South West Scotland	Région de Bruxelles Brandenburg Hamburg Mecklen-Vorpommern Sachsen-Anhalt Thüringen Estonia País Vasco Aragón Alsace Midi-Pyrénées	Auvergne Gelderland Utrecht Noord-Holland Noord-Brabant Bratislavský kraj Länsi-Suom Stockholm Östra Mellansverige Sydsverige	Principado de Asturias Com F. de Navarra Cyprus Groningen Friesland Drenthe Overijssel Flevoland Zeeland Limburg Pohjois-Suomi Åland Mellersta Norrland	
	Medium (22.90 < A < 30.13)	Niedersachsen Nordrhein-Westfalen Cataluña Comunidad Valenciana Andalucía Nord – Pas-de-Calais Prov. Alpes-Côte d'Azur Lombardia	Latvia LithuaniaŚląskie Pomorskie Lisboa Slovenia North West Yorkshire&H East Midlands West Midlands Emilia-Romagna Lazio	Jihovýchod Moravskoslezsko Schleswig-Holstein Galicia Castilla y León Región de Murcia Picardie Haute-Normandie Lorrain	Pays de la Loire Bretagne Poitou-Charentes Aquitaine Languedoc-Roussillon Steiermark North East (UK) Wales Northern Ireland (UK)	Strední Cechy Jihozápad Severovýchod Bremen Saarland La Rioja Basse-Normandie Bourgogne Franche-Comté Limousin Liguria	Prov. Trento Niederösterreich Kärnten Oberösterreich Salzburg Tirol Vorarlberg Itä-Suomi Småland med öarna Norra Mellansverige

Low (A < 22.90)	Veneto Toscana Campania Puglia Isole Sicilia Łódzkie Małopolskie Lubelskie Wielkopolskie Dolnośląskie Continente		Canarias (ES) Piemonte Marche Abruzzo Calabria Sardegna Észak-Alföld Dél-Alföld Podkarpackie Świętokrzyskie	Podlaskie Zachodniopomorskie Opolskie Kujawsko-Pomorskie Warmińsko-Mazurskie Centro (PT) Západné Slovensko Stredné Slovensko Východné Slovensko	Severozápad Strední Morava Castilla-la Mancha Extremadura Illes Balears Champagne-Ardenne Corse Valle d'Aosta Pr. Bolzano Friuli-Venezia Giulia	Basilicata Közép-Dunántúl Nyugat-Dunántúl Dél-Dunántúl Észak-Magyarország Malta Burgenland Lubuskie Algarve Alentejo Reg Aut dos Açores Reg Aut da Madeira Umbria Molise

Source: Runiewicz (2013)
The indicator is indexed by the EU-25 average, and the respective values are given in parenthesis

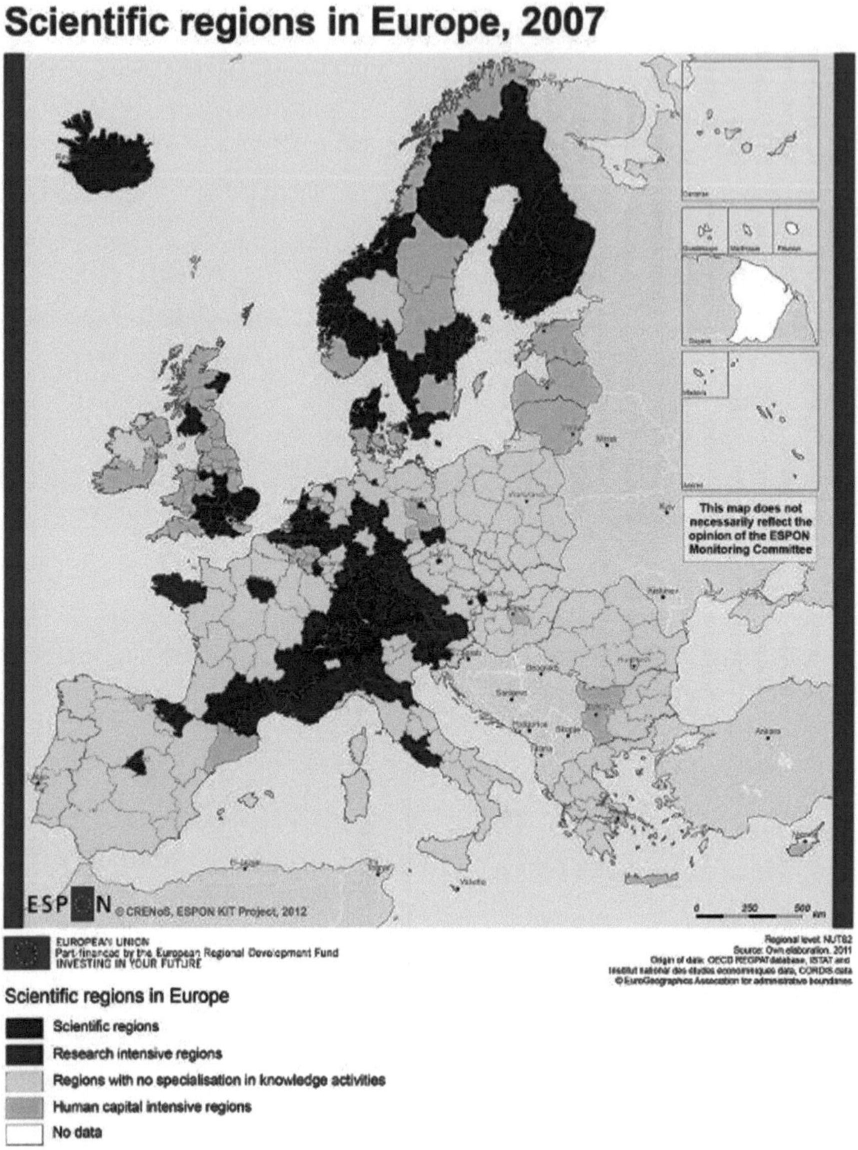

Map 2.6 "Scientific regions" in Europe (Source: KIT (Knowledge, Innovation, Territory) Final report 2013, www.espon.eu)

EU regions – 126 were those with no specialization in knowledge activities. Among the 74 scientific regions, 59 belonged to EU-15 Member States countries and three belonged to the EU-10 group of countries (including Bulgaria and Romania). Strong scientific regions were mostly agglomerated and located in central and northern Europe, namely in Austria, Belgium, Luxembourg, France

(Paris), Germany, Ireland (Dublin), Denmark, Finland, and Sweden, with some notable exceptions in the east, such as Praha, Cyprus, and Estonia, and in the south, such as Lisboa and Attiki. Regions with no specialization in knowledge activities were located mainly on the peripheral territories of Europe, and 'research-intensive regions' were concentrated in territories characterized by a manufacturing productive specialization (Northern Italy, German regions). Finally, as expected, 'human capital-intensive regions' were located mainly in northern Europe.

In general, the KIT project highlighted that knowledge accumulation inside a region also requires networking activity or the acquisition of knowledge from outside. Consequently, scientific regions were also identified as networking regions. Furthermore, the report concluded that a very high number of EU regions, mainly in eastern countries and in the southern peripheral countries were below the EU average in terms of innovation and knowledge-creation capacities.

2.3.3 Knowledge Transfer Capacity: University, Business, and Government R&D Intensities

Technological externalities require efficient innovation systems that connect universities, private enterprises, and government institutions. In some regions such as the EU-10 countries' regions, the innovation system and the underlying technological infrastructure are not sufficiently developed. These regions are generally characterized by relatively low business R&D intensities. For these regions, absorption capacity is embodied mainly in university labs and government research centers. Therefore, transfer capacities and institutional interfaces are necessary conditions for knowledge diffusion within the productive system. Since the direct measures of knowledge transfer among institutional sectors such as higher education, the private business sector, and the government sector are not available, some imperfect proxies, such as R&D intensities could be applied. These measures, along with the regional endowment of knowledge-intensive services (KIS), communication infrastructure, and population density, which enable the carrying of ideas from one individual to another, are discussed below for the EU regions.

Globally, the EU-15 Member States are performing much better in terms of total R&D investments. Finland, Sweden, Germany, Austria, and France belong to the group of leading countries in terms of R&D intensities (with share of total R&D expenditure ranging from 2 % to 3.5 % for 2007 (Eurostat 2011)). Sweden is clearly Europe's best performing country, with an R&D intensity about twice that of the European average.

For new Member States of the EU-10 group of countries, R&D intensities are still relatively low, despite strong positive tendencies during the considered period (ranging from 0.4 % in Cyprus to 1.5 % in the Czech Republic in 2007) (Eurostat 2011). The latter, and smaller states such as Estonia and Slovenia, have caught up significantly. Currently, Slovenia and the Czech Republic outperform some of the EU-15 states (mainly Spain, Portugal, and Italy) in terms of R&D intensity.

Broadly, the business sector realizes about two thirds of total R&D spending on average in the EU. Private sector R&D is considered especially crucial for the innovation and economic growth of regions. It results in the technology that brings new products and services to the market place. However, only a few of the EU Member States have relatively high R&D expenditure performed by the business sector; they are Finland, Sweden, Germany, Austria, and France. The EU-10 states, along with some southern EU-15 states, have relatively lower performance in business R&D. This fact suggests that for new Member States public R&D can still play an important role in the field of knowledge transmission. Government R&D expenditures as a percentage of total R&D is particularly significant for Poland.

In general, Map 2.7 shows that R&D intensities tend to concentrate geographically around capital cities or in big metropolitan areas, where they can benefit from the economies of agglomeration and urbanisation. The Swedish and Finnish regions are clearly the best performing ones, with Västsverige spending almost 5 % on average between 2000 and 2008, followed by Sydsverige (4.3 %), Pohjois-Suomi (4.2 %), Stockholm (4.2 %), and Södra Sverige (4.1 %). The situation is similar for the spatial distribution of business and R&D expenditures. The average value of Swedish regions' business R&D intensity is higher than the ones obtained by most European regions. The latter indicates a higher potential for knowledge creation and diffusion. The situation is slightly different in the field of government R&D intensity. While Sweden clearly outperformed all other European countries in terms of university and business R&D intensities, it holds a weak position in the field of government R&D. The average university, business, and government R&D intensities of the EU regions relative to their country average are illustrated in Figs. 2, 3, 4, 5, 6 and 7 in the Annex 2.

Similarly, the German regions of Baden-Württemberg (4.2 %) and Berlin (3.7 %), followed by the Austrian regions of Wien (3.4 %) and Steiermark (3.4 %), performed very well in terms of average share of R&D expenditure during the considered period. For these regions, absorption capability has been shown to be strong, and their high performance in terms of business R&D intensity suggests an important potential for knowledge creation capacity (see Annex 2).

On average, the potential for knowledge transfer capacities of the southern periphery of the EU is relatively weak. This is especially true for the Greek, Portuguese, and Spanish regions. With the exception of the two capital regions, Madrid and Lisbon, the average levels of R&D intensities for the Greek, Portuguese, and Spanish regions are low and range from 0.3 % to 0.5 % on average for the period 2000–2008. The only region performing relatively well in business R&D is the Cataluña region in Spain (with the number above 1 %).

Among the EU-10's best performing regions are the Czech regions of Střední Čechy (2.5 %) and Praha (2.1 %), followed by Slovenia (1.30 %) and the capital regions of Hungary and Poland – Közép-Magyarország (1.2 %) and Mazowieckie (1 %). The rest of the EU-10 regions spent well below 1 % of their GDP on R&D

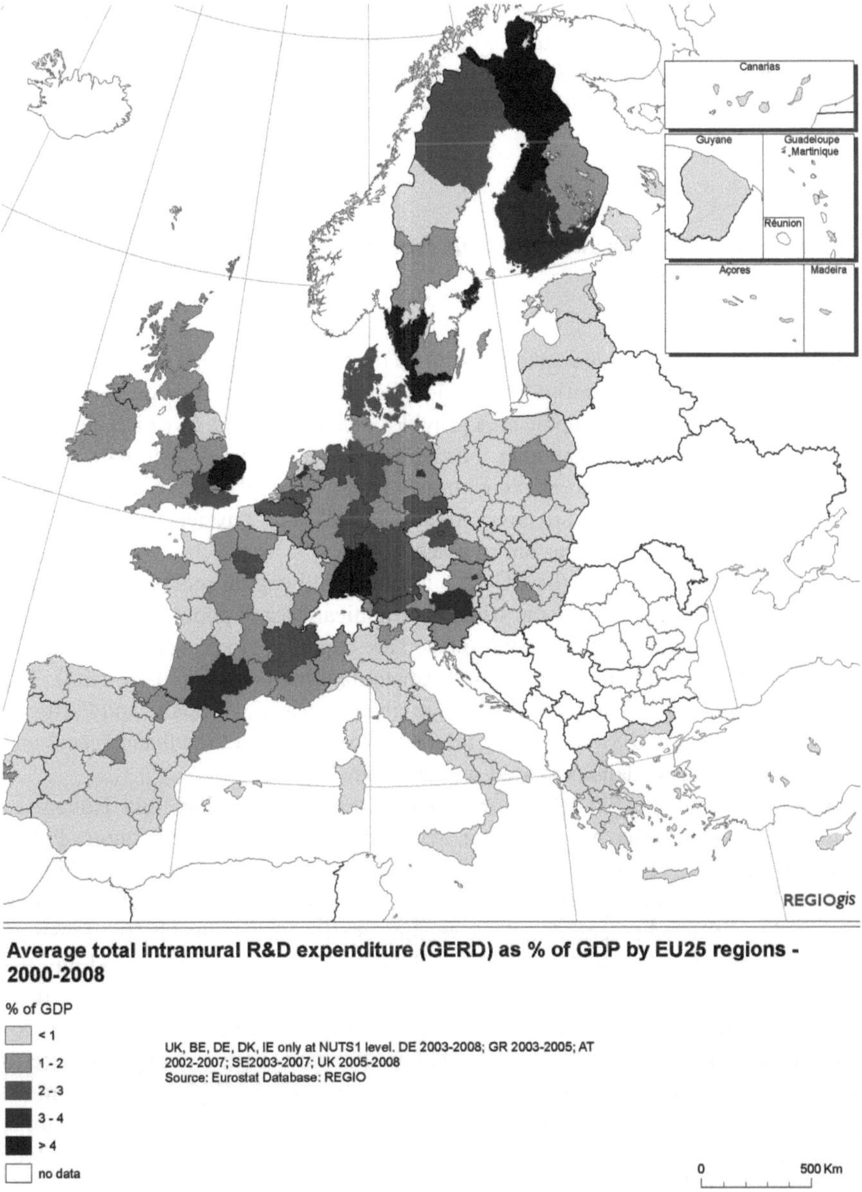

Map 2.7 Average intramural R&D expenditure as percentage of GDP in EU regions (2000–2008) (Source: Eurostat – REGIO)

activities. This is also a common trend for most of the southern regions of the EU-15: Portugal, Spain, Italy, and Greece, with some exceptions for Province Trento in Italy. For many of the EU-10 regions structural funds and public R&S sources have become a significant, if not the main source, of R&D funding.

Given their extremely low levels of R&D intensity, especially in the private business sector, it is questionable whether sufficient resources are devoted to the development of knowledge transmission mechanisms such as private-public interfaces in the above-mentioned regions.

2.3.4 Potential Knowledge and Innovation Diffusion Capacities

Knowledge externalities are considered to be the most compelling in the context of cities. The spatial concentration of individuals, capacities, information, and knowledge within a limited geographic area provides an environment, in which ideas flow quickly from one person to another. Furthermore, since dynamic externalities arise from communication between economic agents, their effects should be more readily observable within an environment where both physical proximity and infrastructure communications are in place.

Table 2.2 illustrates that the EU-15's northern and capital regions have greater potential for knowledge diffusion when communication and population density are taken into consideration. Among those listed are capital regions such as: the Berlin, London, Wien, Stockholm, Madrid, and Amsterdam regions (Noord-Holland). The rest of the EU-15, especially the southern regions such as Italy, Spain, and Portugal, as well as Finland and Sweden, have relatively low levels of broadband access and degree of population density (based on data from 2008).

Only a few of the EU-10 regions, mainly the Czech, Polish, and Hungarian capital regions of Praha, Mazowieckie, and Budapest, along with the bigger Polish agglomerations – Śląskie, Małopolskie, and Dolnośląskie – have the highest knowledge diffusion potential among this group of regions. The differences in knowledge diffusion potential in these regions are also determined by their distinct administrative devolution. The capital of Poland, Warsaw, is incorporated into the Mazowieckie region, whereas Prague and Bratislava are city regions. Given their relatively high broadband access, these regions could have better opportunities for outsourcing and in-sourcing activities. Although there isn't any strong evidence to show how Internet technologies affect innovation processes, Web-based communication technologies such as browsers, Websites, search engines, online forums, email, blogs, and wikis enable easy exchange of information and retrieval of digitalised knowledge content.

There is a small group of regions for which both the level of broadband and the population density are relatively low. This group contains more peripheral regions, predominantly rural or mountain regions of Spain, Italy, and some islands such as Açores or Madeira. For these regions, the broadband infrastructure can only be treated as compensation for their relatively sparsely populated areas. The widespread use of ICT systems and equipment can promote teleworking and lifelong learning and accelerate information diffusion (Runiewicz-Wardyn 2008b).

Table 2.2 Knowledge diffusion potential in the context of high-density regions

Y	X	Population density by NUTS2 (2008)			
		High (X > 195.83)		Medium (85.1 < X < 195.83)	Low (X < 85.1)
Households with broadband access	**High Y > 63**	Vlaams Gewest Noord-Brabant Malta South West Utrecht Hamburg East Midlands Hessen East of England North East Baden-Württe Stockholm Yorkshire&H	Berlin Limburg West Midlands Gelderland Wien South East London North West Noord-Hol Nordrhein-Westfalen Zuid-Hol Com.de Madrid	Luxembourg Denmark Wales Dolnośląskie Mazowieckie Overijssel Niedersachsen Bayern Schleswig-Holstein Groningen Sydsverige	Norra Mellansverige Flevoland Manner-Suomi Åland Länsi-Suomi Småland med öarna Övre Norrland Östra Mellansverige Pohjois-Suomi Zeeland Mellersta Norrland Västsverige Friesland (NL) Etelä-Suomi Scotland Itä-Suomi
	Medium 50 < Y < 63	Région wallonne Bruxelles-Capitale Lisboa Rheinland-Pfalz Illes Balears Małopolskie Cataluña Praha Canarias (ES) País Vasco	Principado de Asturias Slovenia Cantabria Opolskie Wielkopolskie Kujawsko-Pomorskie Sachsen-Anhalt Podkarpackie	Ireland Aragón Algarve Estonia Latvia Lithuania Tirol Comunidad Navarra Podlaskie	Kärnten Südösterreich Mecklenburg-Vorpommern Lubuskie Zachodniopomorskie Salzburg Burgenland Dunántúl Niederösterreich

(continued)

Table 2.2 (continued)

Y	X	Population density by NUTS2 (2008)			
		High (X > 195.83)		Medium (85.1 < X < 195.83)	Low (X < 85.1)
		Śląskie	Oberösterreich		
		Közép-Magyarország	Pomorskie		
			Łódzkie		
			Thüringen		
			Nyugat-Dunántúl		
			Westösterreich		
			Közép-Dunántúl		
	Low Y < 50	Emilia-Romagna	Prov Trento	Severovýchod	Valle d'Aosta
		Veneto	Cyprus	Northern Ireland	Prov Bolzano
		Liguria	Umbria	Reg de Murcia	Reg. Açores (PT)
		Lazio	Abruzzo	Severozápad	Reg. Madeira (PT)
		Lombardia	Friuli-Venezia Giulia	Stredni Morava	Castilla-la Mancha
		Attiki		Észak-Alföld	Extremadura
		Com. Valenciana	Toscana	Brandenburg	Castilla y León
		Sachsen	Marche	Lubelskie	Warmińsko-Mazurskie
		Continente	Piemonte	Észak-Magya	La Rioja
		Centro (PT)	Alentejo	Galicia Andalucía	Dél-Dunántúl
		Norte	Świętokrzyskie		Jihozápad
			Stredni Cechy		Dél-Alföld
			Jihovýchod		Steiermark

Source: own calculations based on data from Eurostat – REGIO

2.4 Regional Typology of Innovative Potential and Technological Capabilities in the EU

On the basis of the different aspects of innovative potential, technological leadership, and growth of the EU regions, which were investigated in the previous sections, this section presents a more integrated view of the EU-25 Member States regions. Table 2.3 reveals the relationship between economic and technological development of the EU regions. All the EU regions were grouped into six different categories of regions according to their average GDP per capita (for 2000–2008 in PPP) and innovative and technological potential.

Regions that belong to the technological and economic leaders' category have a high GDP per capita (2008), a high number of high-tech patent applications per capita, and a high share of employment in the high-tech manufacturing and services sectors. This group includes only the EU-15 group regions, especially their capital regions and big agglomerations such as London, Stockholm, Hamburg, Île de France, Bruxelles, Wien and Groningen, Åland, Utrecht, Bremen, and others. The regions that belong to the category of the innovative, technological, and economic leaders manage their entire innovation process well.

Table 2.3 shows that there aren't any regions meeting the criteria of high GDP per capita and poor innovative and technological performance. This suggests that innovation and particularly technological innovation are key drivers of economic growth. It is through the 'knowledge externalities' that new knowledge quickly becomes social knowledge, acts to enhance productivity at the enterprise level, and contributes to sustained long-term economic prosperity.

Innovation and technological leaders achieving average GDP per capita levels, may have well-developed innovation (R&D intensity) and technological capacities (high-tech patents), but may not be very efficient in valorising their technological achievements. Their important technological advances have been accompanied by relatively lower GDP per capita (2008) in comparison to other EU regions. These regions include the French regions (Alsace, Bretagne, Franche-Comté, Midi-Pyrénées, Provence-Alpes-Côte d'Azur, Rhône-Alpes), the Spanish capital (Madrid), Dutch (Flevoland and Limburg), German (Niederösterreich, Nordrhein-Westfalen, and Rheinhessen-Pfalz), Austrian (Kärnten and Steiermark), Finnish (Länsi-Suomi and Pohjois-Suomi), Swedish (Östra Mellansverige), the UK (South West), and Belgium (Vlaams Gewest). It can be expected that if these regions succeed in improving their efficiencies in exploiting their innovative and technological opportunities to produce higher productivity and added value, they will join the group of "technological and economic leaders."

The next group of regions includes areas with average innovative and technological potential, but high GDP per capita (2008). This group includes the Swedish (Mellersta Norrland, Norra Mellansverige, and Småland med öarna), the Dutch (Groningen and Zeeland), the German (Bremen), the Austrian (Oberösterreich, Vorarlberg, and Salzburg), the Italian (Lazio and Emilia-Romagna), and Luxembourg regions. These could be classified as highly productive, with high

Table 2.3 Regional typology of innovative potential and technological capabilities in the EU

GDP p.c. 2008	High innovative potential and technological leadership		Average innovative potential and technological leadership	Poor innovative potential and technological followership	
High (Y > 28,100)	Åland Baden-Württemberg Bayern Berlin East of London Etelä-Suomi Hamburg Hessen Île de France Lombardia London Luxembourg Manner-Suomi	Noord-Brabant Noord-Holland Norra Mellansverige Övre Norrland Bruxelles South East Stockholm Sydsverige Utrecht Västsverige Wien Zuid-Holland Mell Norrland	Mellersta Norrland Groningen Bremen Groningen Luxembourg Vorarlberg Salzburg Oberösterreich Lazio Norra Mellansverige Emilia-Romagna	Zeeland Småland med öarna	
Medium (20,266 < Y < 28,100)	Alsace Bretagne Madrid Flevoland Franche-Comté Kärnten Länsi-Suomi Limburg Midi-Pyrénées Niederösterreich	Nordrhein-Westfalen Östra Mellansverige Pohjois-Suomi Provence-Alpes-Côte d'Azur Rheinhessen-Pfalz Rhône-Alpes South West Steiermark Vlaams Gewest	Aquitaine Auvergne Basse-Normandie Cataluña Centre (FR) Drenthe East Midlands Friuli-Venezia Giulia Gelderland Haute-Normandie Itä-Suomi L. Roussillon Liguria	Lisboa Niedersachsen Nordrhein-Westfalen North East North West Overijssel P. la Loire Piemonte Provence-Alpes-Côte d'Azur Région Wallonne Sch-Holstein Toscana West Midlands	Aquitaine Aragón Bourgogne Com de Navarra Corse Friesland (NL) Illes Balears La Rioja Limousin Lisboa Lorraine Marche Nord – Pas-de-Calais País Vasco

				Poitou-Charentes
				Umbria
Low (Y < 20,266)	Bratislava	Brandenburg	Notio Aigaio	Lithuania
	Cyprus	Észak-Magyarország	Abruzzo	Łódzkie
	Estonia	Jihovýchod	Alentejo	Lubelskie
	Észak-Alföld	Mecklenburg-Vorpommern	Algarve	Małopolskie
	Jihozápad	Podkarpackie	Andalucía	Molise
	Közép-Dunántúl	Nyugat-Dunántúl	Attiki	Peloponnisos
	Közép-Magyarország	Pomorskie	Basilicata	Podlaskie
	Malta	Reg. Açores	Calabria	Pr. de Asturias
	Moravskoslezsko	Sachsen	Campania	Puglia
	Dél-Dunántúl	Sachsen-Anhalt	Canarias (ES)	Reg. Madeira
	Dél-Alföld	Severovýchod	Cantabria	Región de Murcia
	Praha	Sicilia	Castilla y León	Śląskie
	Severozápad	Stredné Slovensko	Castilla-la Mancha	Sterea Ellada
	Slovenia	Stredni Cechy	Centro (PT)	Świętokrzyskie
	Mazowieckie	Stredni Morava	Comunidad Valenciana	Thessalia
		Thüringen	Dolnośląskie	Východné Slovensko
		Západné Slovensko	Dytiki Ellada	Warm-Mazurskie
			Extremadura	Wielkopolskie
			Galicia	Zachodniopom
			Ipeiros	Latvia Kuj. Pomorskie
			Kentr. Makedonia	

Note: Within each cell, regions are listed in decreasing order of GDP per capita (2008) innovative potential but less focus on high-technology sectors. Although the number of patents is relatively high in these regions, it is less fruitful in terms of high-tech industries.

The regions with average innovative potential and technological leadership and an average level of GDP per capita include the Finnish (Itä-Suomi), the Spanish (Cataluña), the Dutch (Overijssel, Gelderland, and Drenthe), the German (Niedersachsen, Nordrhein-Westfalen, and Schleswig-Holstein), the UK (East Midlands, West Midlands, North East, and North West), the French (Aquitaine, Auvergne, Basse-Normandie, Centre (FR), Haute-Normandie, Languedoc-Roussillon, Pays de la Loire, and Provence-Alpes-Côte d'Azur), the Italian (Friuli-Venezia Giulia, Liguria, Piemonte, and Toscana), the Belgian (Région Wallonne) regions, and the Portuguese capital, the Lisboa area. Regions belonging to this group have a low number of high-tech applications and relatively high knowledge-intensive employment capacities compared to the other EU regions.

Despite the fact that their innovation systems perform weakly, these regions achieve average levels of GDP per capita. They cannot reach the level of wealth obtained by technological and economic "leaders."

The reasons for that can be both outside as well as within their specific local aspects (e.g. lack of presence of high-tech clusters). For the regions belonging to this group, sustained economic development requires a significant strengthening of their technological and innovative bases. It can also be that some regions are actually ready to upgrade their technological capacities and could soon shift into the category of technological or economic leaders.

It is also possible to see that none of the regions belonging to the EU-10 countries is prone to concentrate in the group of either average and high-positioned technological and economic leaders. As indicated in Table 2.3, these regions are characterized by poor wealth creation, despite their high or average technological and innovative capacities. Without exception, all the Hungarian, Czech, Polish, and Slovakian regions belong to this category. Compared to the other EU-10 regions, several capital regions of the above-mentioned countries, such as Közép-Magyarország, Praha, Mazowieckie, and Bratislava, followed by other EU-10 regions (Észak-Alföld, Jihozápad, Közép-Dunántúl, Moravskoslezsko, Dél-Dunántúl, Dél-Alföld, Severozápad) and smaller member countries of the EU-10 (Cyprus, Estonia, Malta, Slovenia) possess high innovative potential and technological leadership. It is expected that these regions, in turn, are first to achieve higher levels of GDP per capita compared to the other EU-10 or southern EU-15 regions.

The regions with average innovative potential and technological leadership, but with poor levels of GDP per capita include the rest of the Hungarian areas (Észak-Magyarország, Nyugat-Dunántúl), the Slovakian regions (Stredné Slovensko and Západné Slovensko), and the Czech (Jihovýchod, Severovýchod, Strední Cechy, and Strední Morava), Polish (Lubuskie, Podkarpackie, and Pomorskie), German (Mecklenburg-Vorpommern, Brandenburg, Sachsen, Sachsen-Anhalt, Thüringen), Italian (Sicily), and Portuguese (Açores) regions. Despite their average performance in terms of innovation and technological capacities, their abilities to turn this potential into sustained economic development is low.

Finally, "innovative and technological followers" and those "staying economically behind" are characterized by low levels of per capita GDP and poor innovative and technological performance. Even though many of these regions are characterized by relatively good absorption capacities, they have not yet reached the capacity to create knowledge and wealth. Despite their common characteristic in terms of underdeveloped capacities, the degree of dispersion within this group is relatively important. This is the biggest group of regions belonging to this category as it includes all of the remaining EU-10 group regions along with the majority of the southern regions of the EU-15. Given that the degree of homogeneity within this category of regions is by far the lowest compared to the other groups, the performance of these regions should be interpreted with caution. For example, this group contains both national economies (like Latvia and Lithuania), and the poorest regions of Greece, where high-tech patenting activity is practically non-existent and their share in R&D intensities are extremely low. Nevertheless, for all these regions, further efforts to strengthen their innovative and technological capacities are necessary to enhance the rate of their economic catch up.

In summary, economic prosperity is related to the innovative and technological advancement of regions. Table 2.3 shows two extremes. It is very rare to encounter a case where a high per capita GDP is accompanied by poor innovative and technological performance. On the other hand, it is very common that the technological and economic "laggers" always go in pairs.

2.5 Spatial Distribution of Technological and Innovative Potential of the EU

2.5.1 Spatial Concentration of High-Tech Industries in the EU

The accelerating pace of technological advances and their diffusion through the use of ICT and growing trends towards offshoring and outsourcing raise several questions regarding the spatial allocation of knowledge-based activity across the EU. More specifically, are the industries that emphasize research and innovation more spatially concentrated? Does high-tech employment exhibit different patterns of geographical concentration than high-tech patent concentration does?

The EU countries' distribution of innovation and knowledge-intensive activities can be evaluated through the use of Gini's concentration coefficient. The Gini index measures spatial concentration based on the Hoover-Balassa index of revealed comparative advantage, and can be written as (Brulhart 2000):

$$HB_k = \frac{x_r{}^k}{\sum_r x_r{}^k} \Big/ \frac{\sum_k x_r{}^k}{\sum_r \sum_k x_r{}^k}$$

For each country the concentration index is calculated on the basis of the regional share of the country's employment or patents in a given sector *k*. The index compares the weight of sector *k* in all the other sectors in the region to the weight of this sector in all sectors at the country level. The indexes are calculated for two time periods for the patent activity (2001–2004 and 2005–2007) and just for one time period for the employment activity between 2005 and 2007. The more geographically concentrated the industry, the higher the Gini value. Purely random patterns of geographical dispersion lead to measures of around 0.3. Therefore, clusters within advanced economies with some mobility of factors and firms between regions should reach Gini values of above 0.3 (Solvell et al. 2003). Table 2.4 shows that knowledge-intensive industries are characterized by strong spatial concentration.

High-tech industries tend to be more concentrated than less knowledge-intense industries, such as industries in which employment is medium high-tech. In general, employment in both types of industries was more concentrated in the smaller EU states than in the larger ones (except Lithuania). The concentrations index in the EU-10 states is higher than in the EU-15 states (except for the Republic of Poland and Lithuania). Among the EU-10 group of countries Malta, Cyprus, Slovakia, Czech Republic, and Slovenia had the highest Gini coefficient concentration in high-tech and medium high-tech employment levels.

High-tech employment is the least concentrated in the southern states of the EU and in more recent EU Member States. The levels of concentration approximate to random geographical dispersion were observed for Spain (0.38), Greece (0.20), and Portugal (0.39). For Lithuania, Poland, and Estonia the Gini coefficients for high-tech employment were lower than 0.6. This relatively low level of spatial concentration in high-tech industry employment can be explained by the small share of high-tech manufacturing in total employment, intra-regional economic disparities within these countries, as well as their administrative devolution (Lithuania and Estonia are considered as single regions at NUTS 2 level). Therefore, the mechanisms based on economies of scale and agglomeration externalities will matter less for these regions.

In all the EU Member States, concentration of high-tech patents is extremely high. Changes in Gini coefficients between 2001–2004 and 2005–2007 were small for every state, implying small changes in concentration; however, there was a slightly increasing trend for Belgium, Finland, Portugal, Slovenia, Malta, and Estonia. Going into details of the spatial concentration of high-tech industry in the EU requires more detailed study of the six basic groups of high technology: aviation; computers and automated business equipment; communication technology; lasers; micro-organisms and genetic engineering; and semiconductors.

The biggest concentration of the *Aerospace industry* (NACE Rev 1.1 codes 35.3 Manufacture of aircraft and spacecraft) is located in three EU countries – Germany, France, and the UK – and accounts for 80 % of the EU's value added, 72 % of

Table 2.4 High-tech patents and employment Gini coefficients in EU Member States in 2005–2007

Country	High-tech employment	Medium high-tech employment	Knowledge-intensive services	High-tech patents		
				2002–2007	2001–2004	2005–2007
Austria	1.12	0.95	0.84	0.92	0.92	0.91
Belgium	0.61	1.01	1.05	0.92	0.90	0.92
Germany	1.47	1.61	1.01	0.91	0.90	0.92
Denmark	0.79	0.93	1.17	0.93	0.92	0.93
Spain	0.38	0.73	0.88	0.93	0.91	0.93
Finland	1.80	0.87	1.32	0.86	0.83	0.88
France	1.00	0.87	1.03	0.90	0.88	0.91
Greece	0.20	0.37	0.58	0.93	0.95	0.90
Italy	1.05	1.15	0.94	0.92	0.92	0.91
Portugal	0.39	0.55	0.61	0.93	0.88	0.94
Sweden	0.81	0.98	1.34	0.89	0.87	0.90
United Kingdom	0.86	0.82	1.13	0.89	0.87	0.90
Estonia	0.96	0.53	0.83	0.87	0.85	0.92
Czech Republic	1.42	1.55	1.08	0.93	0.93	0.92
Cyprus	–	–	0.59	1.09	1.20	0.95
Hungary	1.15	1.08	1.05	0.92	0.90	0.91
Slovakia	1.49	1.40	0.96	0.96	0.95	0.95
Slovenia	1.03	1.44	1.03	0.92	0.86	0.95
Lithuania	0.54	0.35	0.71	0.96	0.96	0.95
Latvia	–	–	0.83	1.45	1.29	1.43
Malta	1.66	0.61	0.90	1.93	3.86	1.43
Poland	0.52	0.83	0.90	0.92	0.90	0.91

Source: own estimations

employment, and 94 % of total R&D spending (Hollanders 2006). Space activities account for less than 10 % of the activities in the aerospace industry. Most innovative activity in the aerospace industry takes place in the following top 20 regions: French (Midi-Pyrénées, Île de France, Provence-Alpes-Côte d'Azur Rhône-Alpes, Haute-Normandie), German (Niedersachsen, Hamburg, Bayern, Baden-Württemberg, Bremen, Nordrhein-Westfalen, Schleswig-Holstein), UK (South West, South East, East of England, West Midlands, East Midlands), Spanish (Com. de Madrid), Swedish (Östra Mellansverige), and Denmark. Meanwhile, the five regions of Midi-Pyrénées, Île de France, Niedersachsen, Hamburg Bayern, and Baden-Württemberg, contribute to the totals by having over 50 % of all patents in this group of regions, with respect to the average number of patents in 2003–2007. As for the EU-10 areas of Cyprus, Lithuania, Közép-Magyarország, and Mazowieckie, they account for the total number of European patent applications of this group. Although the industry's direct economic weight is relatively small, with percentages of 1.0 % in terms of employment, and 1.5 % for value added in terms of total manufacturing (Eurostat 2006), the industry's indirect impact is much more significant. The industry is home to staff with key skills and possesses key technologies in different fields such as electronics, software, telecommunications, materials, and, more recently, market related and managerial fields.

The *Biotechnology industry* (NACE Rev 1.1 codes 24.4 Manufacture of pharmaceuticals, medicinal chemicals and botanical products) employs approximately 100,000 people (year 2006) in total, of which most work in SMEs. However, the exact figures on its contribution to employment or the number of companies involved in biotech-related activity are unknown. Many of these companies are only partially active in biotechnology so not all companies may be covered by the relative statistical surveys. The pharmaceutical industry employs 615,000 people. Around 15 % of the total number of high-tech patent applications relates to the "microorganizm and genetic engineering" (Bio4EU study, JRC/IPTS 2007). The industry is highly research intensive, with almost half of its employees involved in R&D functions; therefore, it relies strongly on interaction with universities and depends heavily on their science-based research.

Geographically speaking, innovative activity in biotechnology fields is dispersed around the EU regions, however it remains clustered around major universities, which specialize in biotechnological research. Broadly, the top 20 regions of the EU-15 and EU-10 account for 61 % and 2 % of EU patent applications in the biotechnology industry, respectively (2007). In terms of the top 20 patenting EU regions, based on the accumulated number of patent applications in the 2003–3007 period, the following regions are taking the lead: German (Bayern, Nordrhein-Westfalen, Baden-Württemberg, Hessen, Berlin, Niedersachsen, and Rheinland-Pfalz), French (Île de France and Rhône-Alpes), Denmark, UK (East of England and South East London), and Belgium (Vlaams Gewest and Rég. Wallonne), Dutch (Zuid-Holland and Zeeland Gelderland), Italian (Lombardia), and the Spanish capital (Comunidad de Madrid). Amongst them Bayern, Nordrhein-Westfalen, Baden-Württemberg, Hessen, Berlin, Île de France, Denmark (Medicon Valley, located around Copenhagen and the Skaane Region), East of England, and Vlaams Gewest regions produce roughly 50 % of all patent applications for this group of EU regions. The best performing EU-10 regions in terms of patenting activity include those that contain capital cities within them, for example Mazowieckie (Warsaw), Praha, Közép-Magyarország (where Budapest is located), Észak-Alföld, and Slovenia. These regions take the first five places of the top 20 ranking (based on the accumulated number of patents in 2003–2007) and produce over 45 % of all patent applications for this group of EU regions.

The *Communication industry* (NACE Rev 1.1 codes 32 Manufacture of radio, television, and communication equipment and apparatus) employs approximately 3.6 million people (as of 2007) in the EU, of which 1.16 million work directly for telecommunication service operators. The four largest member countries (Germany, UK, Italy, and France) account for almost 80 % of the telecommunications value added in the EU. Employment in the communication technology industry in the EU-10 dropped between 1995 and 2004 as a result of the late phase of transition (privatization of big state-owned companies) (Havas 2006). However, since their accession to the EU in 2004 employment increased again, particularly in

Hungary and Poland. These two countries have one of the biggest communication technology clusters in the EU, employing some 10 % and 5 %, respectively, of the total number of people employed in the manufacture of radio, television, and communication equipment (Eurostat 2005).

Geographically, the group of the top 20 patenting regions of the EU-15 in this industry is concentrated in the German (Bayern, Baden-Württemberg, Nordrhein-Westfalen, Niedersachsen, Hessen, and Berlin), French (Ile de France, Bretagne, and Rhône-Alpes), Swedish (Stockholm and Sydsverige), Finnish (Etelä-Suomi and Länsi-Suomi), Dutch (Noord-Brabant and Zeeland), UK (South East, South West, and East of England), Belgian (Vlaams Gewest), and Danish (North Jutland) regions. These regions account for 71 % of the total number of EU patent applications in the communication industry.

The share of the EU-10 in the number of patent applications in the EU within the communication technology industry is very small. The top 20 regions of the EU-10 account for some 2 % of the total number of EU patent applications to EPO in the communication industry. Half of the total number of patent applications for this group of regions is located in Hungarian Közép-Magyarország and the Polish Lubuskie and Podkarpackie regions. The communication industry can potentially play a crucial role in industrial specialization and thus, for regional development in these regions. The Hungarian Közép-Magyarország region significantly outperforms all the other EU-10 regions in terms of patent activity.

The *Computer industry* (NACE Rev 1.1 codes 30 Manufacture of office machinery and computers) employs some 1.6 million people in the EU, of which 140,000 are employed in the EU-10 (9 %). Germany employed almost 26 %, UK 16 %, Italy 10 %, and Ireland 8 % of the total labor force in this industry for 2008. The Czech Republic is the only country from the EU-10 group of members that contributes significantly to employment in this industry – 6 % of total employment (Eurostat 2008). Overall, the computer industry is prospering in the EU. For example, in 2007, the number of people employed in the software industry and computer services was over 51 % higher than it was when compared to 1999. High growth in the number of computer services has taken place, for instance, in Austria, Hungary, Spain, Ireland, The Netherlands, and Slovakia. Large firms (with more than 250 employees) are more common in the manufacture of office machinery and computers (NACE 30), and produce most of the sector's value added and R&D output.

The top 20 regions from the EU-15 and EU-10 groups account for 76 % and 0.8 %, respectively, of the EU-25's total number of patents in this industry. Most of the EU-15's innovation activity in this industry is concentrated in the following top 20 regions: German (Bayern, Baden-Württemberg, Nordrhein-Westfalen, Niedersachsen, Hessen), Swedish (Sydsverige), French (Ile de France and Provence-Alpes-Côte d'Azur, Rhône-Alpes, Bretagne), Finnish (Etelä-Suomi and Länsi-Suomi), Dutch (Noord-Brabant and Zeeland), Belgian (Vlaams Gewest), the UK (South East, East of England, London, South West), and the Italian (Lombardia) regions. The share of new Member States in EU patent applications within computer technology is very small. The top 20 regions from the EU-10 account for 0.8 % of the total number of EU patent applications in the computer industry. Over half of the patent applications for the top 20 EU-10 regions are located in the Hungarian (Közép-Magyarország), Czech (Praha), and Polish (Mazowieckie) regions, as well as in Slovenia, Cyprus and Estonia.

The EU *Semiconductor industry* (NACE Rev 1.1 codes: 32 Manufacturing of radio, television and communication equipment and apparatus) shows strong growth in automotive electronics, industrial and medical equipment, wireless communication, and consumer electronics. In 2008 the industry employed 215,000 workers and contributed to generating approximately 10 % of the EU's value added.[2] The top 20 regions of the EU-15 account for 90 % of all EU patent applications in the semiconductor industry. Most innovation activity in this field is concentrated in the German (Bayern, Baden-Württemberg, Nordrhein-Westfalen, Hessen, Sachsen, Berlin, Rheinland-Pfalz, Thüringen), French (Rhône-Alpes, Île de France, Prov. Alpes-Côte d'Azur), Dutch (Zeeland and Noord-Brabant), UK (South-East and East of England), Austrian (Steiermark), Belgian (Vlaams Gewest), Italian (Lombardia and Sicily), and Finnish (Etelä-Suomi) regions, sites of the largest silicon semiconductor design clusters in Europe.

The top 20 EU-10 regions account for roughly 2 % of EU patent applications in the semiconductor industry. Most of the innovation activity in this industry is done by Poland's Mazowieckie region, the Czech Republic's Jihovýchod and Strední Morava regions, Slovakia's Bratislava region, and smaller states of the EU-10 (such as Estonia, Lithuania, Slovenia, Cyprus, and Malta).

Finally, the *Lasers and optical technologies industry* (NACE Rev 1.1 codes 33 Manufacture of medical, precision and optical instruments, watches and clocks) employs around 300,000 people (2010).[3] In addition, the jobs of more than two million employees in the EU's manufacturing sector depend directly on photonic products. The European photonics industry is dominated by SMEs, which makes the industry both more adaptable to change and more sensitive to international market fluctuations that may take place.

The largest contributors to the EU's value added in precision instruments were Germany (34 %), the UK, France, and Ireland. Among the newer EU Member States, where precision instruments contributed to a more-than-average extent to manufacturing, were two of the Member States that joined in 2004: Malta and Slovenia (the latter in terms of employment).

The top 20 EU-15 regions account for 77 % of EU patent applications in the lasers and optical technologies industry. A major part of innovation activity in this industry is clustered around the German (Bayern, Baden-Württemberg, Nordrhein-Westfalen, Berlin, Rheinland-Pfalz, and Thüringen), French (Île de France, Rhône-Alpes, and Bretagne), UK (South East, Scotland, East of England, and South West), Dutch (Zeeland and Noord-Brabant), Italian (Lombardia), Belgian (Vlaams Gewest), Austrian (Vorarlberg), Danish, and Irish regions.

[2] Sustainable semiconductor manufacturing in Europe – the future of the industry, Position Paper of the European Metalworkers' Federation (EMF), June 2010, http://ec.europa.eu/enterprise/sectors/ict

[3] www.optik-photonik.de

The share of EU-10 Member States in terms of EU patent applications in the communication technology industry is very small. The top 20 EU-10 regions account for some 2 % of the total number of EU patent applications in the lasers and optical technologies industry. Most of the innovative activity for this group of EU countries is located in Lithuania, Slovenia, and within the Hungarian (Dél-Alföld) and Polish (Mazowieckie, Dolnośląskie, Podkarpackie, and Wielkopolskie) regions.

2.6 Summary and Conclusion

Three decades after the accession of the southern Member States to the EU and almost a decade after the accession of the Central and East European countries to the EU, there are still big wealth disparities within and across its Member States, with the former growing faster than the latter ones. Regions with GDP per capita largely above the European average extend from the UK South of England to the Benelux, western Germany, and western Austria, and end in the northern part of Italy. Nonetheless, within these countries, regions with the highest GDP per head over the past years have been Inner London, Brussels, and Luxembourg, followed by Hamburg, Île de France, Wien, Uusimaa, and Stockholm. Most regions belonging to the new Member States as well as the southern European periphery, including the Portuguese, Spanish, southern Italian, and Greek regions, are characterized by relatively low levels of GDP per capita.

A common feature of regions with high GDP per capita is their high technological and innovative potential. The opposite trend in turn was observed in the poorer regions. The latter shows that the technology gap provides a fundamental potentiality for lagging behind regions to catch up. Yet, factual catch up is only possible if the regions lagging behind develop sufficient technological infrastructure to improve knowledge absorption, transfer, and diffusion capacities. In some regions, such as EU-10 countries' regions, RISs are not in place yet. These regions are generally characterized by relatively low business R&D intensities. For these regions, absorption capacity is embodied mainly in university labs and government-led research centers. Capital regions and larger agglomerations have greater potentials for knowledge diffusion because of the relatively better communication infrastructure and population density. It is therefore not surprising that the Bratislava, Közép-Magyarország, Praha, and Mazowieckie regions are among the technological, innovative, and economic leaders amongst the EU-10 group of regions.

From a dynamic point of view, all the EU regions performed well. As a result, the gap between the richest and poorest EU regions has, in fact, narrowed since 2000. In the context of the "technological gap" and endogenous growth theory, it is important to understand what role knowledge spillovers and technological change played during the past decade in the growth and catching up of the EU regions.

Chapter 3
The Efficiency of Regional Innovation Systems (RIS). The Role of High-Tech Industry and Knowledge-Intensive Services

3.1 Introduction

As pointed out in the introduction, the literature on endogenous growth demonstrates that the regional capacity to sustain innovation is embodied in the region's institutions and resources – namely, its underlying technological infrastructure. Moreover, the process of innovation does not follow a straightforward linear model path from the laboratory to the marketplace. In other words, rather than being sequential, the innovation process is a spiral. Schematically, private business research is fuelled by diverse university research and diverse types of expertise coming from customers and suppliers. The latter also feeds back into university research.

Despite the fact that innovation and technological advancement are considered the engines of growth, few attempts have been made to explicitly model the role of technological advancement and the innovation process in the growth of European regions. Innovation and technological advancement are two very complex processes with mutual interdependencies. Efficient innovative systems create a friendly environment for technological advancement, productivity growth, and high-tech manufacturing and services employment. In return, the greater concentration of high-tech employment leads to increasing innovativeness and patent activity.

Therefore, taking into account the potential mutual influence of regional innovation processes and technological advancement, Chaps. 3 and 4 are aimed at empirically testing the efficiencies of EU Member States' regional innovation systems (RIS) and their interlinkages with intra, inter, and extra-regional knowledge spillovers.

The remainder of the following chapter is organized as follows: The first section gives an overview of the subject-related literature. It presents the empirical results of the selected studies and aspects of their methodology. The second section discusses the challenges faced by the innovation systems of the EU-10 regions from a historical perspective; understanding the transformation of their systems of

M. Runiewicz-Wardyn, *Knowledge Flows, Technological Change and Regional Growth in the European Union*, Contributions to Economics,
DOI 10.1007/978-3-319-00342-9_3, © Springer International Publishing Switzerland 2013

innovation and industrial restructuring will enable a proper interpretation of the results of the model regressions and their possible implications for regional policies. The third section presents the results of the estimations and model output and explains them. The respective contributions of university R&D and business R&D to knowledge creation at the European regional level are outlined; the relationship between university R&D and private business R&D in the knowledge creation process in European regions is determined; the role of human qualifications within the process of knowledge creation is defined; and the relative efficiencies of RIS in the EU-15 and EU-10 are identified. The chapter ends with a summary and conclusions of the most important findings.

3.2 Knowledge Transfer Mechanisms and the Concept of RIS in Empirical Studies Literature

3.2.1 Evaluation of RIS: Empirical Studies of the EU Member States

Since the concept of RIS is relatively new, its research framework is still being developed. Generally speaking, however, there two sets of studies applied in the analysis of RIS. The first set is based on comparative empirical studies of a limited number of regions in order to explore desirable criteria upon which systemic innovation at the regional scale might occur. Such comparative analyses of RIS aim to identify not only generalities but also individual features of regions, analyzing new trends in development and the resulting policy implications. The main objectives of these studies can vary from understanding the functioning of RIS, its main factors and mechanisms, to attempting to assess the impact of RIS on regional competitiveness and innovation. Among the most recent are the studies by Greunz (2003); Cooke et al. (2000); Sternberg (2000); Asheim et al. (2003).

One general feature of these studies is that they explore both the general determinants of innovation potential and the innovative linkages in RIS within a limited number of EU regions.

In the study by Cooke et al. (2000), the authors explore the key organization and institutional dimensions of RIS, based on a sample of 11 EU regions (Baden-Württemberg, Wallonia, Brabant, Tampere, Centro, Féjer, Lower Silesia, Basque country, Friuli, Styria, and Wales). One of the study's key contributions was its highly detailed analysis of regions' innovation performance and potential. In general, Wallonia (Belgium), Friuli (Italy), Styria (Austria), Tampere (Finland), and the Brabant (Holland) regions had well-functioning RIS, with the Tampere and Styria regions serving as good examples of well-developed systems of science parks and business incubation. The Portugal Centro region, however, did not demonstrate this characteristic. A similar conclusion was made with regard to almost all the regions of the Central and Eastern Europe (CEE) countries, which, according to the authors of the study, failed to fit the criteria of the existence of RIS. Despite this, one of the four successful examples of RIS in this group of regions included the Polish Lower Silesia (Dolnoslaskie) and Hungarian Fejer regions. With regard to the former, the authors concluded that, based on its universities, the region had potential but little regional systemic capability. Conversely, the Fejer region had a well-developed regional innovative culture in automotive and electronic engineering but had no government or research base.

Another study by Asheim et al. (2003), aimed to draw on the SME and regional dimension of innovation policies, based on a sample of 13 Nordic regions (Oslo, Stockholm, Helsinki, Gothenburg, Malmö/Lund, Aalborg, Stavanger, Linköping, Jyväskyla, Horten, Jaeren, Salling, and the Icelandic regions). The study explored the existence of similarities and differences between regional clusters of SMEs in the different Nordic regions. In the context of the Nordic cluster, initiatives on social networking arrangements proved to be a particularly successful way of boosting and securing social capital and trust. In addition, it was concluded that SMEs that draw mainly on an analytical knowledge base and which innovate through science-driven R&D (e.g. in biotechnology) tend to collaborate with global partners in the search for new and unique knowledge. Whereas, SMEs that draw mainly on a synthetic knowledge base and which innovate through engineering-based user-producer learning, tend to collaborate more with regional partners.

The study by Sternberg (2000), which aimed to assess the qualitative and quantitative determinants of innovation potential of 11 EU regions, varied from large agglomerations (with the country's capital) to more or less peripheral areas (such as Vienna, Stockholm, Barcelona, Alsace, Baden, Lower Saxony, Gironde, South Holland, Saxony, Slovenia, and South Wales). One of the most interesting outcomes of the study was the suggestion that focusing on a single region without accounting for its geographic, historical, and general macroeconomic conditions is not a sufficiently accurate approach to explain RIS performance. Based on the German Saxony case study, the study concludes that the region still suffers from the consequences of the economic and political transformation process resulting from the country's reunification. The asymmetry in innovative pressures in eastern and western German regions explains why Saxony and other eastern German regions are unable to catch up with the western German regions in the development of innovative products and processes (for more see Fritsch and Schwirten 1999). Another interesting outcome concerns the three metropolitan regions, namely Stockholm, Vienna, and Barcelona. These regions have been the most dynamic in their respective countries in recent history. However, the potential for regional innovation is currently underexploited, since numerous businesses have transferred their R&D facilities abroad.

The second set of studies includes a case-by-case analysis of individual RIS and provides important insights into the nature and dynamics of regional development. Such case studies can help to identify the main drivers of the emergence, growth, and maintenance of RIS. The individual case studies enable the researcher to map the various kinds of interactions among the RIS actors, social networks, and institutions that support innovation activity on a regional scale (Cumbers et al. 2003; Cooke et al. 2002, 2000). Despite this, although the detailed analysis of the case studies reflects the unique features of every RIS, there is no single model from which generalizations (and, thereby, specific policy recommendations) can be made on the dynamics of a successful RIS.

Moreover, the observation of a phenomenon in the case of one region does not mean it occurs in another region and, therefore, comparative case studies allow for a more thorough investigation. Nonetheless, such comparative studies have only limited implications for economic theory and, as such, the development of RIS theory.

The selected literature on RIS provides some evidence of the relationships between the processes of learning, networking, and innovation based on the experience of particular groups of regions. However, the institutional dimension of the interactions of major actors in RIS has not been sufficiently explored. The literature has almost ignored the experience of the regions of the ten most recently acceded EU Member States, both in empirical quantitative and comparative case studies analysis. Consequently, the validity of their recommendations for regional innovation policy is somewhat questionable.

Thus, in order to understand the interlinkages between RIS in the EU, further research must be undertaken which considers not only the elements and their role in the absorption and diffusion of knowledge, but also the relationships between these elements and the wider socio-economic environment. Previous studies on RIS provide clues to understanding regional development and its institutional environments; however, they are not conclusive and based largely on a limited number of successful examples from the EU-15. Likewise, case study analysis enables the identification of some special features but can hardly be applicable to a broader level of policy and theory. The common institutional and systemic problems of the EU-10 require this group of regions to be considered separately to the total sample of EU regions. Doing this would allow proper policy conclusions to be drawn, thus identifying the current stage of RIS development in the EU-10.

3.2.2 Evaluating the Interlinkages and Efficiency of RISs in EU

Knowledge creation within RIS is a complex process, characterized by various looping and feedback relations between its actors. A number of studies exist which focus on particular aspects of the knowledge-creation process within RIS. Most of them lead to the concept of the “knowledge production function” (KPF), a concept originally Griliches (1979) introduced as a tool to measure an innovative input–output relationship. Examples of such regional applications of the KPF approach include studies by Jaffe (1989), Anselin et al. (1997); Feldman and

Audretsch (1999), and Acs (2002). Although most of the literature concerns US regions, the methodology of Griliches (1979) and Jaffe (1989) has been tested and improved by many European authors: Italy (Capello 2001), France (Autant-Bernard 2001), Germany (Fritsch 2002), and at the European regional level (Greunz 2003; Guellec and Van Pottelsberghe 2003; Dunnewijk et al. 2008, and Wintjes and Hollanders 2011).

In general, these studies show a strong association between innovation inputs that have been captured by regional investment in R&D and regional innovative performance that has been captured by patents. In this section, only the most common methodology is presented.

Jaffe (1989) was the first to consider a simultaneous model of influence. He was also the first to use the Griliches (1979) production function framework for a sample of 29 US states over a disconnected period of 8 years. In Jaffe's analysis, corporate patents are the outcome of university and business R&D expenditures. University R&D depends on both a set of structural variables and business R&D. The latter, in turn, is a function of university R&D. Jaffe's findings suggest that both industry and university research are important sources of innovation for the business sector. Furthermore, universities encourage the location of industry R&D spending in nearby areas. However, in neither his global model nor that of different technical areas, does he find evidence of a significant impact of industrial R&D on university R&D.

Inspired by the work of Jaffe (1989), Feldman and Florida (1994) investigated a very similar model but, instead of corporate patents, they consider commercial product-innovation citations as a measure of innovative output. Moreover, they include among their independent variables the concentration of firms active in related manufacturing industries and specialized business services. The authors provide evidence that both industrial and university R&D positively and significantly influence private business innovation. Furthermore, the regression outcome indicated that university R&D fuels industrial R&D. However, contrary to the findings of Jaffe (1989), their estimates show that the reverse also occurs, with private business R&D positively and significantly influencing university R&D.

Anselin et al. (1997) took one step further in the cross-section database by testing the regression for a sample of 125 US metropolitan areas. A major development in comparison to other studies in this field is the introduction of spatially lagged explanatory variables. Although the estimated elasticities vary substantially according to the specifications, like Jaffe (1989) and Feldman and Florida (1994), they found evidence of the positive impact of university R&D and industrial R&D on innovative output. Moreover, university R&D significantly influences private business R&D. Like Jaffe (1989), but contrary to Feldman and Florida (1994), they found no evidence of a significant contribution of private business R&D to university research.

In terms of studies on EU regions, the results of Guellec and Van Pottelsberghe (2003) suggest that direct government funding of R&D performed by firms has a positive effect on business-financed R&D (this is unless the funding is targeted towards defence activities). Defence research performed in public laboratories and universities crowds out private R&D, whereas civilian public research is neutral for business R&D. The stimulating effect of government funding varies depending on its generosity. In most cases, it increases up to a certain threshold (about 10 % of business R&D) and then, beyond that threshold, decreases.

Based on the aforementioned model, Greunz (2003) tested an extended sample of 153 European regions and suggested the following: the patenting activity in European regions depends on both private business and university R&D. Since university R&D is essentially concerned with the generation of fundamental scientific knowledge, its impact on patent applications is less and it takes more time compared to private business R&D, which is more market oriented. The relationship between university R&D and business R&D is characterized by feedback relations. University R&D and business R&D positively influence each other through various channels of knowledge transmission. While private business R&D depends on the region's endowment of human capital with medium and high levels of qualification, only human capital with a high level of qualification positively influences university R&D.

A knowledge production factor analysis by Dunnewijk et al. (2008) demonstrates that, at the regional level, high scores on patents and business R&D are associated with a large share of high-tech manufacturing. Conversely, Grimpe and Patuelli (2011) prove that both public and private R&D are relevant for nanomaterial patenting in an interactive way. They suggest that, from 2000 to 2004, co-located R&D provided opportunities for knowledge spillovers and collaboration between actors in the relevant German regions.

Inspired by the above-mentioned research on EU regions and the earlier results from the US, this chapter focuses on the European regional landscape for which this kind of investigation has never been undertaken. We pay particular attention to the qualification level of human capital (both tertiary and advanced), since the latter has been shown to determine to a large extent both a region's absorption capacity and a company's choice of location (Malecki 1985; Galbraith and De Noble 1988; Cockburn and Henderson 1998).

3.3 Innovation Systems of the EU-10 Member States Considered from an Historical Perspective

Differences in innovation performance across EU regions result from differences in the socio-political environments of the two major groups of the regions. After all, innovation processes take time and therefore it seems natural to consider the long-term historical perspective when characterizing such systems. It is important to underline here that, during the past two decades, the EU-10 regions observed three important changes that affected their innovation systems: the transition of innovation systems; administrative decentralization; and the increasing role of clustering and localized knowledge flows (LKS), especially in knowledge-intensive sectors.

Considering their GDP levels, the EU-10 Member States have, in general, inherited quite large R&D systems. This is why they have good research and development capacities in terms of university and laboratory infrastructure (see Radosevic and Auriol 1999). However, these capacities are nonetheless very concentrated and thus regionally imbalanced. The dismantlement of the centrally planned and controlled system found many of the regions unprepared to cope with the large-scale restructuring that followed. In addition, the emphasis of transition policies on short-term efficiency has reduced the scope for redistribution. The role of regional policy is being downgraded and its funds marginalised. Regional authorities have, in essence, continued to be agents of the central government, but with no meaningful decision-making authority. The process of transition increased centralization at the territorial level and limited self-government to the municipal level (Gorzelak 1996). As a result, most of the EU-10 countries entered into the transformation with a very weak regional base for R&D institutions and with a reduced capacity for independent decision making among the regional authorities (Radosevic 2000).

Furthermore, R&D capacities were not directly related to the needs of the local productive capacities. One reason for this was the dominance of inter-regional vertical production chains (and a few horizontal intra-regional linkages), coupled with the social role of large firms in the local community (Radosevic 2000). As a result, socialist countries were deprived of agglomeration economies and, instead, arose from a diversified regional economic structure. All of this had an enormous influence on the long-term capacity of these countries to rebuild their regional competitive advantages (Grabher 1997, p. 111).

Since the beginning of the transition process, economic agents in the post-socialist countries were often forced to change their technological paths in order to attain competitiveness on the national and international markets (Fritsch and Werker 1999). Such changes in the technological path led to a reduction in the amount of resources devoted to innovation activities and a depreciation of the knowledge stock[1] (Gokhberg 1994; Jasinski 1994). This caused a considerable brain drain of scientific personnel, who moved to other spheres of activity or to other countries, thus reducing the knowledge stock available in regions and necessary for the generation of innovation. (Nesvetailov and Artyukhin 1995). The role of capital cities represents the highest potential for RIS. For example, 77 % of research personnel in Hungary are located in the Budapest region (EUROSTAT 2012). The national framework in the CEE also plays an important role in determining inter-regional disparities. As a result, most of the EU-10 regions (except for the relatively better positioned capital cities) entered the transition period with a

[1] In addition, some regions were occupied predominantly by agriculture employment and, since the agricultural market sector and technological opportunities were weak, the potential for growth in these regions was relatively low.

very poor knowledge of absorption and transfer capacities and the dominant role of the public sector in regional R&D financing.

In addition, the lack of knowledge networking and spillovers at the regional level hindered the remedying of the shortage of knowledge by potential intra- or extra-regional sources of R&D. Potential intra-regional linkages were not used as a source of efficiency improvements or innovation. Instead, the dominant linkages were inter-regional and organized within individual sectors that were dependent on the center (Gorzelak 1996; Radosevic 2000).

All the above factors, in addition to the lack of local R&D institutions and external finance (due to an undeveloped banking system and lack of venture capital), created a totally different starting position for these regions and, for this reason, they require separate consideration in further research.

The transition of the innovation systems in the EU-10 Member States induced a number of specific challenges at the regional level that resulted from profound reorganizations of the division of labor within and between R&D organizations. The transition also encouraged privatization, which produces significant differences in the reconstitution and restructuring of enterprises and 'narrows' the national R&D infrastructure in support of innovations. It should also be mentioned that the EU-10 regions have not developed appropriate informal institutions that increase the efficiency of RIS. These problems have led to severe differences with regard to the generation of knowledge and innovation between the two groups of EU regions.

At the micro level, liberal trade and FDI from the advanced EU-15 countries provided the EU-10's domestic firms with technological, market, and managerial knowledge, and thus enriched the regional knowledge stock. For this group of countries (behind the technology frontier) technology transfer has been very important for growth and closing the gap with the EU-15 countries. Imported goods and FDI inflows, mostly from advanced EU-15 member states served as channels of international knowledge spillovers and enabled EU-10 firms to rebuild competitive advantages, especially in the automotive and electronics industries (Bellak and Narula 2008; Runiewicz 2005; Umiński 2002).

In sum, the past 10 years have seen localized activities foster regional innovation and network alignment through the establishment of innovation centers, S&T parks, free economic zones, business incubators, and regional development agencies. In this respect, it could be argued that regional policy in the CEE has embraced the main elements of the RIS framework. Most often these initiatives are initiated through foreign assistance and government programmes.[2] This allows us to conclude that the EU-10 regions follow the neo-institutional (or neo-corporatist, Viale et al. 1998) model of RIS linkages, where the role of government is very unlike

[2] As a result, they are not always understood by the relevant business actors. For example, very often S&T parks that were originally meant to strengthen knowledge exchange operate instead as renting areas (Webster 1996).

many of the EU-15 regions that moved towards a more liberal, neo-evolutional[3] model of RIS. The evolutional theory assumes that RIS "constantly transforms, propelled by self-perpetuating local or regional processes" (Szlachta and Zaucha 2012, p. 3). Even though their innovation systems are being operationalized through interlinkages between universities, industry, and governments, their regions have not got enough power to endogenously generate universities-industry-government interactions. Thus, the government is expected to play a more important role in coordinating innovation processes within RIS.

3.4 Framework and Model

Chapter 2 highlighted the low innovation capacities of the EU-10 regions and clearly indicated their weak performance in terms of R&D intensities and patent activity. This weak performance is despite the fact that all EU-10 regions are not necessarily disadvantaged in terms of qualifications of human capital compared to other European regions. Furthermore, this group of regions had to undergo major changes, such as the transition of innovation systems, administrative decentralization, and the emergence of cluster policy. As a consequence, their RIS are not yet, or only partially, in place.

In this context it is interesting to investigate simultaneous equation models, paying special attention firstly to the EU regions as a whole and, secondly, to the part of EU-10 regions that are characterized by extremely low levels of business and university R&D investments. This information allows the systemic properties of RIS to be assessed for all EU regions and, in particular, the EU-10. It also allows the researcher to derive expectations about their relative performance. Comparing these results with the actual level of regional innovation activity described in Chap. 2 may lead to conclusions about the relative efficiency of RIS and the importance of systemic differences.

Building upon aspects of new economy theory, the following section aims to identify the characteristics of LKS and their influence on regional innovation processes. The method of estimation is based on the KPF, a framework initiated by Griliches (1979) and first implemented at the aggregate level by Jaffe (1989).

The application of KPF as a method to evaluate RIS has several limitations. It offers only a partial picture of the actors, structure, and dynamics of RIS and their influence on regional competitiveness. It also overlooks many other elements comprising the institutional structure of RIS such as the institutional set-up of

[3] The evolutionary interpretation, in contrast, assumes that, within specific local contexts, universities, government and industry are learning to encourage economic growth through the development of "generative relationships" and a spontaneous convergence of the industrial and academic worlds (Leydesdorff and Etzkowitz 1997). The role of the government in the evolutionary interpretation of the Helix Model is limited to cases of so-called "market failure." Many of the EU-15 regions in the UK, Germany, and the Netherlands, discussed in the previous section, could meet the criteria of the neo-revolutionary model of RIS.

training, the innovative culture, the organization of firms, inter-firm relationships, and the system's financial sector. Moreover, it does not consider, for example, cultural factors, such as the eagerness of actors to learn and interact (Cooke et al. 1997). In spite of these shortcomings, the KPF is an efficient model to investigate a region's technical efficiency, gaining the maximum possible output from a given set of inputs. This addresses the major difference in results across EU regions, namely the differences in the level of inputs devoted to innovation and the efficiency with which they were used in the knowledge production process.

According to Griliches (1979), innovative input is best reflected by new knowledge, which is primarily embodied in R&D efforts. However, the present model extends beyond this by incorporating a number of additional variables in the model. The model estimates a cross-sectional knowledge production of 186 geographic regions in the EU in the period 1998–2008, based mainly on NUTS II territorial units. The use of this classification for regional analysis presents an important limitation in the research. This is because regions defined within NUTS II do not necessarily correspond to sufficiently homogenous and self-contained regions in the broad sense (Evangelista et al. 2002 (Italy); Doloreux 2002 (Sweden). However, NUTS II is the only classification system used by Eurostat on which territorial data is developed.

$$\text{R\&D output} = \text{f (R\&D input)} \tag{3.1}$$

The model adopts a general version of the Cobb-Douglas production function, which does not impose any restriction regarding returns to scale, where *a* is a constant, while *b* measures the elasticity of knowledge output P with respect to input.

$$\text{R\&D output} = \text{a (R\&D input)}^{\text{b}}\text{exp}^{\varepsilon} \tag{3.2}$$

Taking the natural logarithm of each side leads to Eq. 3.3:

$$\ln\text{(R\&D output)} = \ln\ \text{a} + \text{b}\ \ln\text{(R\&D input)} + \varepsilon \tag{3.3}$$

To compare the extent to which business and university R&D induces patenting and innovations in the EU regions, the following model is proposed:

$$\begin{aligned}\ln(\text{P}) = {} & \ln\ \text{a1} + \text{a2}\ \ln\ (\text{BR\&D}) + \text{a3}\ \ln(\text{UR\&D}) + \text{a4}\ \ln(\text{HQ}) \\ & + \text{a5}\ \ln(\text{HKIS}) + \text{a6}\ \ln(\text{GDP per capita}) + \varepsilon 1\end{aligned} \tag{3.4}$$

Following Jaffe (1989), the potential interaction between university and business R&D is captured by extending the base Eq. 3.4 to capture the effects of: business and university R&D expenditures; the qualification level of the working age population; and the number of students at the second stage of tertiary education on patent activity. The structure of the knowledge productive system is a proxy for the share of high and medium-high technology industry in total employment. It also aims to describe the application of innovation output and the state of the knowledge economy. Despite the increasing importance of the service sector in the technological progress of recent years, the manufacturing sector provides the basis for innovation.

High-tech knowledge-intensive services (HKIS) act as the enablers of a knowledge economy since this subgroup includes indicators such as ICT, researchers in scientific disciplines, and employment in S&T. The importance of HKIS to local innovative activity may also be a proxy for the increasing role of offshoring and outsourcing high-skill jobs to foreign nations (a more realistic extent of offshoring and outsourcing can be hard to measure). Finally, figures on advanced qualifications and employment in high-tech sectors are also expected to offer information on the competence level and talent of a region.

It is assumed that, based on regional patent data and GDP per capita discussed in Chap. 2, there is a strong relationship between the number of patent applications per capita and GDP per capita. In other words, richer regions invest more in patenting activity. The GDP per capita is also a basic economic performance indicator of a region and illustrates the region's competitiveness.

As far as innovative output is concerned, the only available innovation measure at the EU NUTS2 regional level is a proxy for patent applications to the EPO. Even if patent data is not a perfect proxy for innovations (as has been already discussed in Chap. 1), there is a strong link between patents and inventions (Griliches 1979; Guelleck and Van Pottelsberghe de La Potterie 1999; Acs et al. 2002).

As suggested by the literature on the organization of a RIS, there is a systemic, two-way interaction between its three major institutional subgroups, namely university-business, government-business, and university-government. Therefore, Eq. 3.4 is further extended by the two additional Eqs. 3.5 and 3.6. University-business relations refer to the amount of investments and research collaboration, whereas government-business relations measure public funding for innovation, technology grants, and R&D linkages.

Considering that the innovative process takes time, next to "year-by-year" regressions are regressions that allow for time lags. Therefore, different combinations of lagged R&D variables were introduced to the system of equations.

The introduction of the two additional Eqs. 3.5 and 3.6 solves the problem of the simultaneity of dependent and independent variables in the model of KPF. For example, not all independent variables in the model are independent from other variables, e.g. previously university-business R&D synergies.

In order to reduce the possible inconsistency of the model, the following system of equations is estimated:

$$\begin{aligned}\ln(P) = \ln a_1 + a_2 \ln(BR\&Dt) + a_3 \ln(UR\&Dt) + a_4 \ln(HQ) \\ + a_5 \ln(HKIS) + a_6 \ln(GDP\ per\ capita) + \varepsilon 1\end{aligned} \tag{3.4}$$

$$\begin{aligned}\ln(BR\&D) = \ln b_1 + b_2 \ln(UR\&D\ t) + b_3 \ln(GR\&D\ t) + b_4 \ln(S) \\ + b_5 \ln(HQ) + \varepsilon 2\end{aligned} \tag{3.5}$$

$$\begin{aligned}\ln(UR\&D) = \ln c_1 + c_2 \ln(BR\&D\ t) + c_3 \ln(GR\&D\ t) + c_4 \ln(HQ) \\ + c_5 \ln(S) + \varepsilon 3\end{aligned} \tag{3.6}$$

Where:

P patent applications to the EPO by priority year at the regional level per million inhabitants (1998–2008);

BR&D total intramural business R&D expenditure at the regional level as a percentage of GDP (1996–2008);

UR&D total intramural university R&D expenditure at the regional level as a percentage of GDP (1996–2008);

GR&D total intramural government R&D expenditure at the regional level as a percentage of GDP (1996–2008);

HQ students at ISCED levels 5–6 as a percentage of all pupils and students at regional level (1998–2008);

GDP per capita is a variable measuring the economic conditions of the regions (1998–2008);

HKIS high-tech knowledge-intensive services as a percentage share of total services (1998–2008);

S employment in high-tech and medium high-tech as a percentage of total manufacturing (DG, DK, DL, DM34, DM35.2, DM35.3, DM35.4, DM35.5) (Source Eurostat – REGIO);

Equation 3.5 explains that business R&D is a function of: university and government R&D expenditures; regional specialization in high and medium high-tech manufacturing (S); and tertiary education human capital (HQ). In previous sections of this chapter we suggested that university R&D fuels industry R&D through mechanisms of technology transfer and diffusion. Educational attainment levels are also particularly important for innovation activities, as the academic literature suggests. The complexity of new technologies and the process of adaptation and implementation of technological know-how are strongly linked to human skills and qualifications. The variables (HQ) indicate creativity and the state of the knowledge economy. Concentrating solely on R&D indicators leads to a situation where the science-based approach is over-emphasized when evaluating the creation of new knowledge and innovation. After all, knowledge is created not only in R&D laboratories, but also in other parts of firms and society as a whole. Yet, the greater the number of qualified human capital, the greater the knowledge externalities they can create.

Finally, Eq. 3.6 states that university R&D depends on: business and government R&D expenditures and the share of human capital with a tertiary qualification and an advanced level of research qualifications. Since the innovation process is created by multiple feedback loops, private business R&D is supposed to fuel university R&D (Kline and Rosenberg 1987). Government R&D is another determinant of university R&D. Governments not only run research institutes and finance higher education, but also subsidise research and development initiatives. Part of government subsidies for R&D are intended to promote high-risk university R&D projects and acquire new expertise, something which, in turn, should give rise to further research. The introduction of the variable of educational attainments in the equation is motivated by the same arguments as those given for Eq. 3.5.

Unfortunately, with such a large number of EU regions it is almost impossible to have complete data. Missing data are therefore important to consider, because they may lead to substantial biases in analyses. Before going to modelling the study of the missing value analysis, which contains the univariate, statistics and corresponding number of missing cases for all the variables have been done. In general, it can be assumed that having up to 20 % of missing data shows the good fit of the regression. In the present model, for variable lnP there is 12 % missing data, which is acceptable. On average, data gaps are estimated at a little more than 22 %, which is also acceptable, although some data would need to be supplemented. That concerns, in particular, lnBRD, lnGRD, and lnURD. For most regions, official R&D data are available for at least the beginning year, the middle year, and the end year of the period 1997–2007. The EU-25 refers to all the current EU Member State regions, except the two latest Member States: Bulgaria and Romania. The regions are analyzed on the NUTS2 regional level, with a few exceptions: namely Ireland, Denmark, and Germany, for which data were only available in the national and NUTS1.

3.5 Estimates Results

Globally, the results of the simultaneous system of equations indicate a high level of similarity between the estimates obtained by the different methods. However, if the estimated coefficients obtained by the two estimation methods are statistically similar, it is relatively difficult to choose one over the other. Klein (1974, p. 150) and Gujarati (1995, p. 679), adopted the 3SLS method, arguing that "single equation methods, in the context of a simultaneous system, may be less sensitive to specification errors in the sense that those parts of the system that are correctly specified may not be affected appreciably by errors in specification in another part."

The estimations are performed by means of a panel data estimation method that allows for random and fixed effects and assume a lag structure for the business, university, and government R&D expenditure at the regional levels. The estimation results for the entire sample of EU-25 and EU-10 regions are presented in Tables 3.1 and 3.2.

As far as the patenting equation is concerned, for the entire sample of EU-25 regions, it is positively and significantly influenced by the following explanatory variables: the university and business R&D expenditures (with the best fitting 3-year time lag); the high-tech knowledge-intensive services (HKIS) (measured as a share of total services); and the overall economic conditions of the regions (measured as GDP per capita).

The results of the estimates indicate that the elasticity of patenting activity for the EU Member States regions with respect to university R&D expenditures is relatively significant and important. Roughly speaking, a 1 % increase of university R&D investment per capita generates a 0.72 % increase of patent applications per capita with a time lag of 3 years (with 99 % significance). With regard to the

Table 3.1 Results of the simultaneous system of equations for EU regions (Estimated with 3SLS)

	Without fixed effects		With fixed effects	
Variables	Coeff.	T-stat.	Coeff.	T-stat.
Log patents				
logBR&D_lag3	0.583	(11.14)***	−0.004	(−0.04)
logUR&D_lag3	0.715	(8.94)***	0.041	(0.27)
LogHQ	−0.795	(−5.58)***	1.191	(2.97)***
LogHKIS	0.996	(8.68)***	−0.171	(−1.14)
logGDP per capita	0.649	(6.41)***	−0.576	(−2.53)**
_cons	−2.408	(−2.04)**	0.062	(2.39)**
Log business R&D				
logUR&D_lag3	0.583	(8.53)***	0.205	(3.41)***
logGR&D_lag3	−0.020	(−0.48)	0.032	(0.83)
logS	23.282	(9.45)***	2.172	(1.22)
logHQ	−0.737	(−5.50)***	0.046	(0.33)
_cons	1.246	(2.68)***	0.070	(6.88)***
Log university R&D				
logBR&D_lag3	0.155	(6.90)***	0.046	(1.27)
logGR&D_lag3	0.082	(3.63)***	0.013	(0.39)
logS	−2.728	(−1.84)*	2.322	(1.48)
logHQ	0.809	(11.59)***	0.189	(1.56)
_cons	−2.893	(−12.79)***	0.050	(5.43)***
N	488.000		488.000	
r2	0.705		0.020	

Source: Runiewicz-Wardyn (2013)

Note: The estimates used the Stata command "reg3." For fixed effects estimations, from original variables their deviations from the mean values within the region were calculated (i.e. the fixed effects estimation). Heteroskedasticity-consistent t-statistics in brackets

Standard errors in parentheses: ***$p < 0.01$, **$p < 0.05$, *$p < 0.1$

estimates for the EU-10 regions, a 1 % increase of university R&D investment per capita generates a 0.44 % increase of patent applications per capita with a time lag of 3 years (with 95 % significance) (Table 3.2).

With respect to business R&D expenditures, the elasticity of patenting activity for the EU regions is relatively significant and important. Broadly speaking, a 1 % increase of business R&D investment per capita generates a 0.58 % increase of patent applications per capita with the best fitting time lag of 3 years (with 99 % significance).

Regarding the estimates for the EU-10 regions, a 1 % increase of business R&D investment per capita generates a 0.32 % increase in patent applications, if introduced to the best fitting time lag of 3 years (with 95 % significance). Compared to the effects recorded for the entire sample, this impact is significantly weaker. This means a cluster-specific factor such as business R&D efforts results in a lower level of efficiency of a business sector in driving innovations in these regions. In other words, these findings may suggest that most of the EU-10 firms are adopters and importers of technology rather than producers of innovations.

Table 3.2 Results of the simultaneous system of equations for EU-10 Member States regions (Simultaneous equation estimates with 3SLS)

	Without fixed effects		With fixed effects	
Variables	Coeff.	T-stat.	Coeff.	T-stat.
Log patents				
logBR&D_lag3	0.318	(2.04)**	0.176	(0.54)
logUR&D_lag3	0.439	(2.54)**	0.489	(1.37)
logHQ	−0.610	(−1.51)	2.054	(1.86)*
logHKIS	1.014	(2.55)**	−0.566	(−1.12)
logGDP per capita	0.619	(1.37)	−0.766	(−1.15)
_cons	−3.613	(−0.98)	0.200	(2.03)**
Log business R&D				
logUR&D_lag3	−0.102	(−0.75)	−0.037	(−0.27)
logGR&D_lag3	0.361	(2.98)***	0.124	(1.34)
logS	7.589	(1.58)	2.687	(0.69)
logHQ	−0.380	(−1.17)	0.283	(1.01)
_cons	0.114	(0.10)	0.026	(0.85)
Log university R&D				
logBR&D_lag3	−0.125	(−1.72)*	0.072	(0.87)
logGR&D_lag3	0.170	(1.96)*	−0.116	(−1.91)*
logS	3.304	(0.99)	−3.390	(−1.34)
logHQ	0.565	(2.56)**	0.286	(1.79)*
_cons	−2.964	(−3.98)***	0.097	(5.02)***
N	93.000		93.000	
r2	0.396		0.145	

Source: Runiewicz-Wardyn (2013)

Note: The estimates used the Stata command "reg3." For fixed effects estimations, from original variables their deviations from the mean values within the region were calculated (i.e. the fixed effects estimation). Heteroskedasticity-consistent t-statistics in brackets

Standard errors in parentheses: ***$p < 0.01$, **$p < 0.05$, *$p < 0.1$

The effect of the 3-year time lag of university and business R&D expenditures on patent applications, as reported in the estimates results, is similar to the results of the studies by Fritsch and Slavtchev (2005) (for the German regions).

However, it is a slightly bigger time lag than that in the results of Kondo (2005) (from 1.5 to 1.7 years), but less than that in the results of Greunz (2003) (4 years).

The time lag related to the impact of R&D investments on patenting may differ from country to country or from discipline to discipline, and depends on formal and informal steps that must be taken to file the patent application. The lagged effect of university R&D on the patent application process may also point to a more basic characteristic of research run by university laboratories. That is to say, any university R&D project has to pass from the stage of basic research to that of applied research and experimental development before it becomes commercially interesting. As a consequence, it could be that the university laboratories focus their R&D activity more on those industries that operate on the basis of applied research knowledge, such as medical or laser technology science, as opposed to biotechnology and semiconductors. The purpose of applied research is to transfer knowledge into practical use, and therefore the concrete R&D solutions to actual problems are expected as soon as possible. Finally, these findings may also suggest delays in the knowledge transfer capacities of local R&D systems. This may result from complex intellectual property transfer policy rules that may slow down knowledge and technology transfers in universities.

3.5.1 Results for Business, Government, and University R&D Efforts

The estimates indicate that universities' R&D expenditures have a significant and positive impact on private business R&D efforts. Broadly speaking, a 1 % increase of university R&D expenditure generates a 0.58 % increase in business R&D expenditures, with a 3-year time lag (if both fixed and random effects are considered). Thus, the results suggest that co-location of universities and firms in the same region clearly matter in the exploitation of knowledge spillovers, especially in industries in which industry R&D, university research, and skilled labor are important inputs.

This outcome is in accordance with other studies by Jaffe (1989); Acs et al. (1992, 1994); Zucker et al. (1994); Anselin et al. (1997, 2000); and, more recently, Greunz (2003), thus providing some evidence of the positive university-industry knowledge spillovers. Furthermore, the findings suggest both university knowledge spillovers (UR&D) and human capital (HQ) development constitute important "locational attractors" for private business sector R&D and for high-technology production (Malecki 1991).

The lagged knowledge spillovers from university laboratories to the private sector emphasize the weak knowledge absorption capacity of the industrial side, such as the companies' competences to use and further develop the new knowledge in question. Firms that invest heavily in R&D are likely to possess a higher absorptive capacity (Cohen and Levinthal 1990) and, as a consequence, a greater number of collaborative R&D projects. The lag effect of university R&D on patent applications may also result from cultural differences or an unwillingness to disclose the outcomes of R&D. It may also happen that, in some sectors, universities lag behind industry in terms of R&D facilities or competences, since their faculty and graduate students tend to ignore recent industry developments.

Furthermore, with all-time lags considered, the estimates indicate that university R&D efforts are a negative and insignificant function of business R&D efforts in the EU-10 regions. This could be explained by the insufficient technological capacities of their firms to absorb knowledge developed elsewhere, or an ineffective transfer mechanism of knowledge from university to the private business sector. The causes of this may lie in the significant lack of industrial incubators and industrial parks, in addition to low levels of formal and informal knowledge flows (R&D agreements, faculty consulting in industry, etc.). Last, but not least, it could reveal the poor innovative culture among university professors and students in the EU-10 regions.

Regarding the impact of government R&D on private business R&D efforts for the entire sample of EU regions, the former appears to have a negative and insignificant influence on the latter. This finding contradicts those of Greunz (2003), in the sample of 153 European regions between 1989 and 1996, where it was found that government R&D appears to significantly influence private business R&D, the elasticity being comparable to that of university R&D.

In view of the systemic evolutionary interpretation of the Helix model, it could be suggested that the EU-25 regions have developed generative relationships and a spontaneous convergence of the industrial and academic worlds (Leydesdorff and Etzkowitz 1997). In fact, in many of the EU-15 regions, the ratio of government-funded R&D to GDP decreased constantly during the 1980s and 1990s. Instead, increases in the intensity of business sector R&D have been driven largely by domestic industry funding. However, the absence of a significant impact of government R&D on the region's patenting activity does not mean it has no impact at all. Government R&D may positively influence business sector R&D by improving its R&D absorptive and transfer capacities (for example, by establishing publicly funded lifelong learning and professional excellence institutions and by improving education infrastructure). In fact, the university R&D variable could capture the effects of government R&D. After all, universities carry out a share of public research by means of grants and contracts.

The role of government R&D efforts seems to be critical for business sector R&D in the EU-10 regions. Broadly speaking, a 1 % increase in government R&D expenditure generates a 0.36 % increase in business R&D expenditures, with a 3-year time lag (with 99 % significance). The latter results show that public expenditures for R&D are an important factor of innovation activity in the above EU regions, especially when considering the low levels of business and university R&D expenditures. For these regions, private sector R&D could have been associated in a stronger way with performance and efficiency in public administration and a country's overall economic performance. The importance of government R&D for the EU-10 regions has also been emphasized in other studies. Krammer (2009) found that public R&D efforts are critical for the late-coming regions in their attempts to catch up with the technological leaders. In view of the Helix Model, it can be concluded that most of the EU-10 regions follow the neo-institutional model of university-industry-government interlinkages.

The estimates of the equation clearly indicate a significant and positive interaction between university R&D and business R&D for the entire group of EU regions. Generally, a 1 % increase of business R&D expenditure generates a 0.15 % increase in university R&D in the entire group of EU regions (EU-25) (if a 3-year time lag is considered). This result is lower than that obtained by Feldman and Florida (1994) for the case of the US (0.22).

The model estimates for the EU-10 regions indicate that an increase in business R&D does not positively influence university R&D. The results could be explained by the fact that innovation systems are not yet in place in these regions, especially in terms of linkages between the industry sector and universities and the relatively high dependence of university R&D in these regions on government R&D funds. Even though the possible causes of such poor university-business R&D interlinkages in the EU-10 were discussed in the previous section, this result must be interpreted with caution and considered in the light of the socio-political context of this group of regions. As explained earlier in this chapter, the entire innovation process is not yet in place in these regions. This is particularly true for the feedback relations between the private business sector and university R&D. The increase in knowledge flows between the business sector and universities could be conditional on public R&D support or EU funds in these regions. For example, Olechnicka conducted a survey on 35 Polish companies, which showed that most of the business-university co-operation involved participation in 6FP of EU. The latter especially concerned high-tech industries, such as aerospace and IT (Olenchnicka 2012).

Government R&D also contributes to fostering university R&D. In fact, for the EU Member States regions, an increase of government R&D expenditure by 1 % resulted in a 0.08 % increase of university R&D with a 3-year time lag (with 99 % significance). This outcome looks slightly better for the EU-10 regions, with an increase of 0.17 % in 5 years (with 90 % significance). This result should not be surprising, since (as mentioned earlier in the section) in these regions the university R&D variable also captures the effects of government R&D. Since university R&D units are funded largely by public resources through grants and contracts, the universities' R&D activity could act as a policy-making instrument for the government.

3.5.2 *Results for the Role of High-Tech Industry and Knowledge-Intensive Services*

Table 3.1 clearly indicates that private business R&D activities are positively driven by high-tech employment (S). The statistically positive effects of this model were obtained only for the entire sample. Broadly speaking, the increase of high-tech employment (S) by 1 % generates an increase of 2.3 % in business R&D. This may also suggest that high-tech industries clustering in a RIS are linked to the existence of intra-regional knowledge spillovers, provided that knowledge spillovers related to R&D activity have the classic characteristic of a non-rivalrous good and cannot be appropriated entirely.

The relationship between R&D and high-tech industries is mutually reinforcing. On the one hand, the presence of high-tech industries increases R&D intensity; on the other hand, R&D efforts affect the growth of high-tech sectors through the process of innovation and technological change. This mutually influencing relationship puts pressure on regions and locations to specialize in more high-tech and knowledge-intensive activities. Generally speaking, the estimate findings are in line with other studies which explore the link between innovation, technological change, and job creation, namely those of Freeman et al. (1982); Freeman and Soete (1994); Vivarelli and Pianta (2000); Edquist et al. (2001); and, more recently, Audretsch et al. (2009).

The results for the EU-10 regions turned out to be positive but insignificant. This could be explained by the relatively small high and medium high-tech employment concentration in the EU-10 regions, which means that the technological and research opportunities of these regions are weaker.[4] The share of R&D expenditure devoted to innovativeness shows that firms strive to narrow the technological gap as fast as possible, and with the use of outside material technology (Gaczek and Stryjakiewicz 2008). According to several studies, R&D departments of companies with foreign capital mainly generated the growing demand for highly skilled labor in the high and medium (Inzelt 2003; Radosevic 2002). This signifies that high-tech clusters and innovative potential proxied by business R&D remain underdeveloped in the EU-10 regions.

The results also show the positive influence the high and medium high-tech-industries sector has on university R&D. This is not surprising, since the growth of high-tech industries has increased demand for skilled human capital, contract R&D research, and academic consulting. This is especially true for industries for which basic research and technology transfer efforts of universities are essential, e.g. biotechnology. High technology start-ups choose to locate in highly innovative regions, as these firms expect to benefit from knowledge spillovers (Audretsch and Keilbach 2007; Anselin et al. 1997). Having said this, university R&D efforts nonetheless play a crucial role in the process of high-technology firm birth. However, more detailed conclusions on the role of the externalities of high-tech firms on university R&D efforts and knowledge flows cannot be drawn without a detailed analysis of the high-tech industry structure, the types of knowledge involved in the university-business transfer, geographical and technological proximities of the university and business agents, and so on.

Furthermore, Tables 3.1 and 3.2 clearly indicate that private business R&D activities are positively driven by high-tech knowledge-intensive services (HKIS), both for the entire sample of the EU-25, and the EU-10 separately. Broadly speaking, an increase of HKIS by 1 % generates a similar increase of business R&D in both groups of regions.

The HKIS, including ICT and R&D-related activities, act as a knowledge diffusion channel within and beyond the home region. They work as a knowledge

[4] with Hungary and the Czech Republic being above average and Lithuania and Latvia having very low levels of employment concentration in both high-tech and business R&D efforts (Lengyel and Cadil 2009).

diffusion channel among companies in local, regional, and global dimensions. In this sense, the presence of HKIS can serve as a proxy for the potential effect of outsourcing. In general, the wider the business R&D network, the greater the capacity to create, capture, and accumulate knowledge in a region and reinforce the cumulative dynamics of innovation. Many authors have agreed that, although ICTs (particularly the Internet) improve the possibilities for remote communication, they do not considerably facilitate the transfer of knowledge (Leamer and Storper 2001; Savy 1998). However, considering the fact that the biggest part of HKIS is concentrated in the regions with the highest R&D and innovative capacities (Chap. 2) one should assume that they have some role to play in knowledge spillovers and knowledge outsourcing.

The high share of HKIS in most innovative EU regions could be explained also by their strong ability to absorb and assimilate foreign-made technology (Benhabib and Spiegel 1994; Borensztein et al. 1998; Cohen and Levinthal 1989; Griffith et al. 2003). In fact, a number of relatively recent studies evidence the rising channel of technology flows due to HKIS, in which R&D outsourcing is an important part (Quinn 2000; Metters and Verma 2008; Lai et al. 2009). These studies also argue that successful flows of knowledge – relating to outsourcing – into home regions depends on the level of financial resources for conducting R&D activities in the home region and the stock of resources available (e.g. qualified human capital, scientists, and engineers). Conversely, the evidence of Turlea and Bogdanowicz (2007) is not in accordance with the general findings of the above studies. The authors show that outsourcing may crowd out domestic R&D efforts and LKS, as is the case in Estonia's ICT-related industry, which is produced on the basis of subcontracting by many Scandinavian firms as part of a common Scandinavian ICT cluster. As a result, most R&D is realized abroad, leaving Estonian investment in R&D very low and maintaining the country's status as a lower added-value performer in the value chain.

In sum, even if broadly, the findings of the estimates point towards a positive relationship between the HKIS, as a proxy for the potential effect of outsourcing, and domestic business R&D intensities, these results must be interpreted with the necessary caution.

3.5.3 The Role of Education in Knowledge Production

It is generally assumed education should positively influence patent activity in regions. However, the results of the estimations in the overall sample of EU regions show that there is a negative relationship between the number of tertiary education students and the number of patent applications. More specifically, an increase in the number of tertiary education students decreases the number of patent applications by 0.79 %. Similarly, the same relationship tested on the EU-10 regions turned out to be negative but insignificant.

This result is somewhat surprising and should be interpreted with caution, since previous studies have emphasized that education is an important determinant of a region's absorption capacity (Cockburn and Henderson 1998) and cluster formations (Malecki 1985; Galbraith and De Noble 1988). The results suggest there could be a negative or minimal influence of the formation of human capital (which is a major part of the lifelong learning process) on innovative output. Furthermore, the results for the selected sample of EU regions show a significant but negative relationship between tertiary education (HQ) and private business R&D efforts. A percentage increase in the number of tertiary education students in the entire group of EU regions generates a decrease of business R&D by 0.74 %. This explains that, while university R&D efforts depend on the region's endowment with tertiary education graduates, business R&D efforts may require more specific and highly qualified graduates.

The above-mentioned model outcome could be explained by the fact that innovation and R&D activities usually involve people with advanced research skills and not tertiary qualification-level graduates. A similar reasoning applies to the EU-10 regions; however, the results turned out to be insignificant for this group.

The number of students with tertiary education (HQ) significantly and positively influences university R&D expenditure. The results confirm that qualified students are a prerequisite for basic research mainly conducted by universities. For the entire group of the EU regions a 1 % increase in tertiary education students (HQ) generates a growth of university R&D by 0.083 %. The influence of the number of students with tertiary education on university R&D efforts seems to be less pronounced for the EU-10 regions, despite the fact that all the EU-10 regions have a relatively high share of tertiary education-based human capital compared to other EU-15 regions. The poor research capacities of higher education institutions in the EU-10 to attract domestic and foreign graduates will result in an eventual brain drain and a diminishing pool of candidates for future academic research and scientific careers.

3.6 Summary and Conclusions

In a nutshell, the patenting activity of European regions depends on business and university R&D, the share of high-tech knowledge-intensive services and the overall economic conditions of the regions (measured as GDP per capita). The relation between university R&D and business R&D is characterized by positive feedback relations for the entire sample of EU-25 regions, e.g. university R&D and business R&D positively influence each other through various channels of knowledge transmission. However, business R&D does not positively influence university R&D activity in the EU-10 regions, which could be explained by the fact that RISs are not yet in place in these regions. Whereas the role of government R&D efforts seems to be critical for business sector R&D in the latter regions. It can be concluded that knowledge flows within their RISs could be conditional on public

R&D support and could still follow the neo-institutional model of university-industry-government interlinkages, unlike many of the EU-15 regions (for which government funding for business R&D turned out to be insignificant) that have developed generative R&D relationships and a spontaneous convergence of the industrial and academic worlds.

Furthermore, the knowledge economy, driven primarily by technological advancement and competitive pressures, increases the demand for high-tech knowledge-based services (HKIS) and industrial clustering. The estimate results for the entire sample of EU regions clearly indicate that university and private business R&D activities are positively linked to high and medium high-tech industries clustering and intra-regional knowledge spillovers. In the EU-10 regions, however, innovative potential proxied by business-university knowledge flows (e.g. *start-ups* and *spin-offs*) and industrial clustering remain underdeveloped. Meanwhile, the role of HKIS as a knowledge diffusion channel seems to be especially important, which suggests the importance of the EU-10 regions' regional and global business R&D networks and outsourcing activities.

Furthermore, while university R&D efforts depend on a region's endowment of tertiary education graduates, business R&D efforts may require graduates who are more specialized and possess advanced qualifications. It also raises the question of what skills and scientific profiles companies expect from students upon graduation from universities and other educational institutions.

In addition, the results of the model show that regions' general macroeconomic conditions (measured by GDP per capita) have a positive and significant impact on the number of patentable inventions. Broadly, this confirms that regional economic conditions do play an important role in explaining the differences in the innovative performance of EU Member States regions.

Chapter 4
The Role of Intra-, Inter-, and Extra-Regional Sources of Knowledge Spillovers. The Evidence of High-Tech Industries

4.1 Introduction

While Chap. 3 focused on RIS and local or intra-regional drivers of the regional innovation processes, this chapter aims to investigate inter- and extra-regional knowledge spillovers across EU regions. Inspired by the previously discussed literature in Chap. 1 and by factual innovation trends in the EU presented in Chap. 2, this chapter aims to investigate geographically and technologically determined knowledge spillovers among European regions and thus contribute to the understanding of the diffusion processes of innovation and knowledge in space. The inflow of R&D spillovers from extra- and inter-regional sources may be an important additional source of technological advantage for many regions.

Inspired by the concept of endogenously determined technological change, and in particular by the works of Griliches (1979) and Jaffe (1989) and by theories of "new economic geography," this chapter focuses on further investigation of the European regional landscape in the context of knowledge creation and intra-, inter-, and extra-regional spillovers in general, and high-tech industries particularly.

With the advanced regional integration process going on in the EU, internalizing productive systems and improving ICT infrastructures of knowledge spreading over various geographical areas may become an important driving force behind technological change across the EU regions. Evaluating inter-linkages only within the local/regional dimension does not portray a complete picture of interactions in an RIS. In fact, many studies show that the R&D activities of surrounding regions and of those with which a common technological profile is shared contribute significantly to the concerned region's knowledge creation. Moreover, scientific and technological knowledge does not diffuse evenly in space, and there may be substantial regional and sectoral differences in the adoption of new technology.

Until now there has been very little evidence of geographically and technologically mediated knowledge flows in the enlarged EU, especially in the context of the EU-10, where high-tech clusters are just evolving. Inter- and extra-regional knowledge spillovers may become an important additional source of

M. Runiewicz-Wardyn, *Knowledge Flows, Technological Change and Regional Growth in the European Union*, Contributions to Economics,
DOI 10.1007/978-3-319-00342-9_4,

technological advance for these regions. Understanding the spatial and technological patterns of knowledge spillovers in the EU-10 regions is highly relevant to the EU's innovation policies and contributes to the theoretical discussion in the field of "new economic geography," and will be discussed in the following chapter.

The remainder of this chapter is organized into five sections, the first two of which review the subject-related literature. Section three discusses the role of EU FPs in knowledge spillovers and R&D networks. In the fourth section the estimated KPF is discussed along with the data, role of variables, and the estimated results. Section five discusses the interview results with the selected scientists in high-tech fields. The chapter ends with a summary and a set of important conclusions.

4.2 Geographical and Technological Proximity in the Empirical Studies Literature

4.2.1 Knowledge Spillovers in the EU and US Regions: Evidence Using Patent and R&D Data

The empirical literature on agglomerations and socio-economic clustering largely supports the importance of geographical proximity for knowledge spillovers (Jaffe et al. 1993; Saxenian 1994; Zucker et al. 1998; Porter 2003; Glaeser 2000). Most empirical studies use patents or labor mobility as measures of knowledge flows. Broadly, they conclude that geographically concentrated industries benefit most from the exchange of knowledge among themselves and should therefore be more innovative and grow at a more rapid pace. The geographic proximity of individual bodies possessing high levels of human capital, skills, expertise, or creative capabilities enables interactions between them, and these interactions facilitate the spillovers necessary for innovation. Furthermore, studies by Jaffe et al. (1993), Knudsen et al. (2007), and Sonn and Storper (2003) found that inventors have a tendency to cite local patents at the local geographical level.

Relatively fewer studies suggest the importance of inter-regional knowledge spillovers. They include the works of Jaffe et al. (1993), Jaffe and Trajtenberg (1996), Anselin et al. (1997), Acs et al. (2002), Smith (1999), and Beal and Gimeno (2002) for the US regions, and Harabi (1997), Maurseth and Verspagen (2002), and Piergiovanni and Santarelli (2001) for the European regions.

Using US patent citation statistics, Jaffe et al. (1993) and Jaffe and Trajtenberg (1996) found that intra-national and intra-state citations occur more often than one may expect from the distribution of patenting activity. To put it differently, patents are more often cited when the citing and the cited patents belong to the same geographical area. Using a spatial econometrics approach, Anselin et al. (1997) and Acs et al. (2002) find clear evidence for knowledge spillovers among neighboring US metropolitan statistical areas (MSAs) in the field of university research. An alternative is to use data on individual interactions that might indicate the presence of knowledge spillovers. Jaffe et al. (1993) pioneer this method, using patent citation data to track flows of information. After controlling the existing distribution of inventive activities, they find that inventors tend to cite other patents from within the same geographical unit, be it a country, state, or Metropolitan Statistical Area. Yet, although this represents a considerable advance, it still uses panel data and therefore cannot illustrate temporal change in the light of proximity. For the EU regions, Maurseth and Verspagen (2002), using patent citations data, show that knowledge flows are larger within single countries rather than between regions located in separate countries, as well as within regions sharing the same language but not necessarily belonging to the same country.

A number of studies focus on long-distance partnerships. In general, they tend to assume that the development of ICT, accelerated technological advance, and competitive pressure to further reduce R&D costs and encourage long-distance knowledge flows, as proposed by Castells (1996) and Cairncross (1997). Frenken et al. (2009) confirm this trend in a survey revealing an overall increase in the number of long-distance partnerships. Ponds et al. (2007) show that this has a significant impact in cases of collaboration between partners from different fields of activity. Singh (2005) notes that the impact of these effects is weaker when the anteriority of collaborations is taken into account. That trend is supported by Almeida and Kogut (1999) and Grossetti and Bes (2003), who show that the effects of physical proximity are a result of previous social relations between local partners, which Autant-Bernard et al.'s work also confirms (2007).

However, there have also been authors who do not believe in space as the ultimate factor and determinant influencing knowledge spillovers. Breschi and Lissoni (2001, 2003), based on their review of recent studies, offer a critical discussion on 'Marshallian externalities' and write (..) that the role of geographical distance in the economics of knowledge transmission is still rather controversial (Breschi and Lissoni 2001, p. 976). However, the authors do not provide any specific evidence denying such knowledge flows.

Others like Piergiovanni and Santarelli (2001), Harabi (1997), Sjöholm (1996, 1997), and Maurseth and Verspagen (2002) suggest that business R&D follow their own path of knowledge spillovers. Piergiovanni and Santarelli (2001) evaluated patent activities in regions in France and examined how such activities related to corporate and university expenditures in R&D in those areas. Other spillovers-related studies also include the work of Harabi (1997) who investigated the effectiveness of various R&D channels of spillovers at the intra-industry level, suggesting that a firm's own investment in R&D is the most important driving force for spillovers. An investigation of a Swedish case (Sjöholm 1996, 1997) indicates that citations to patents originating from more distant countries are less frequent than citations to patents from neighboring countries.

Apart from geographical proximity, it is assumed that technological proximity between regions influences innovative activity. The general idea behind the concept of technological proximity is that in order to generate knowledge spillovers regions must display similar technological profiles. At a corporate level of firms and R&D institutions, the capacity to take advantage of another firm's R&D department's or lab's stock of knowledge depends heavily on the extent of their mutual technological similarity. Since a region's productive system consists of firms and R&D institutions operating in different technological fields, spillovers are expected to be higher between regions with similar technological profiles. The empirical literature about European regions by Paci and Usai (1999) and Autant-Bernard (2006) gives evidence of complementary R&D activities within a region itself but also when collaboration with neighboring regions occurs. Maggioni et al. (2007) also draw the same conclusion for inter-regional collaborations, stating that the scale of "co-patenting activity of two regions is positively correlated to the degree of technological similarity of their innovation systems." Hoekman et al. (2010) provide evidence for the positive effect of technological proximity, which can be summed up as follows: "The larger the differences in research portfolios, the less co-publication activities occur," otherwise researchers of similar technological fields publish in the same journals and operate in the same professional organizations. Similarly, Ponds et al. (2007) claim that the importance of the proximity factor differs for each scientific field and reveals that it ostensibly changes between scientific fields linked to biology and those linked to physics. The results of Ponds' work show that technological proximity and partners' scientific and technological profiles would significantly facilitate inter-individual as well as inter-regional collaborations.

To summarize, these studies show that new knowledge is not evenly distributed throughout society, and knowledge spillovers are determined by technological interests and subject-specific experiences of the partners involved. Studies regarding geographic proximity give very inconclusive evidence and suggest that even though local sources of knowledge spillovers are important, their intensity is less profuse when different estimation methods and different fields of activity are considered.

4.2.2 Knowledge Spillovers in the EU and US Regions: Distance and Travel Time Proximity Metrics

The starting point for a distinction between geographical and technological proximity is that a national economy can be divided into functional urban regions that consist of one or several locations. The borders between functional urban regions are characterized by a decline in the intensity of economic interaction and can be related to labor market regions. Johansson et al. (2002) refer to the different categories of accessibility based on the above interactions and labor mobility: intra-, inter-, and extra-regional. In the EU territorial context of NUTS 3, the regions would only act as approximations of functional regions. However, since the statistical data for these functional regions is scarce, they are often grouped according to the NUTS 2 regions. Even though the NUTS 2 regions are too large to capture the above-mentioned interactions, Dijkstra et al. (2011) argue that this does not undermine the credibility of research studies based on innovation or competitiveness measures as they still capture regional variations in terms of their attractiveness for firms and residents.

When accessibility measures are to be considered, proximity's factual value is discounted in a way that reflects the propensity of economic actors with different locations to travel to different destinations inside and outside of the regions (von Hippel 1998; Karlsson and Manduchi 2001). The following literature distinguishes between two types of proximity – geographic and technological (Maggioni et al. 2007; Hoekman et al. 2010; Paci and Usai 1999; Autant-Bernard 2006; Grossetti and Bes 2003; Singh 2005; Castells 1996; Cairncross 1997). The first is linked to physical distance separating the partners that are based in different regions, whereas the second refers to the idea that knowledge spillovers between two regions may be determined by the degree of technological similarity between their RISs.

Although the related literature reflects just a general agreement on the importance of knowledge spillovers due to geographical and/or technological proximity within or between sectors of activity or within a given country or region, a few attempts have been undertaken to explicitly and fully test inter-regional spillovers at the European level under the KPF framework.

Several studies have examined the empirical measures of the geographical spatial dimension of the transmission of knowledge; however, according to Döring and Schnellenbach (2006) "no consensus is reached about the spatial range that can be attributed to knowledge spillovers, and in fact the majority of studies refuse to quantify the range at all" (p. 384). Most of these studies apply two possible measures of knowledge spillovers: physical distance or travel time. The first measure refers to geographical distance by establishing the distance between two regions (geographical or technological neighbors), whereas the second refers to the time it actually takes for the knowledge to reach the other region.

The calculations of accessibility are based on actual travel times within and between different regions. This implies that improved transport infrastructure can increase the number of opportunities for interaction. In fact, the study by Parent and Riou (2005) on the effects of knowledge spillovers on patent growth rates across 335 EU regions between 1989 and 1999 shows that the quality of transport connectivity and possibly other types of communication connectivity are important factors affecting the transmission of knowledge.

The spatial extent (distance) of knowledge flows has been extensively tested on the US landscape, and one of most valuable studies on the subject was conducted by Acs (2002). The author compares the results of a number of studies on knowledge spillovers across regions using various estimation techniques, and concludes that university research spills over a range of 50 miles from the innovative Metropolitan Statistical Areas (MSAs), while spillovers from private R&D tend to be contained within the MSAs themselves.

Johnson et al. (2002b) investigated physical distance itself rather than economic or innovation effects on proximity. They define distance as a simple scalar value of the way that has to be travelled between two points. Thus, in their model, 2,000 miles is always twice the distance of 1,000 miles. With today's transportation and communication, however, 1,000 and 2,000 miles are not significantly different in terms of journey time as they are within a day's flying distance, and the total travel time difference is not enormous. Moreover, the economic and innovation benefits of proximity are not directly proportional to distance but instead, when plotted on a graph, resemble the shape of a Gaussian curve. Being within walking distance of each other is significantly different from being slightly further away. Likewise, partners within driving distance can exchange information face to face at least once a day but people even slightly further away may not be able to afford the costs associated with travelling to locations involving an overnight stay. Also, the authors suggest that the perception and interpretation of the term "distance" may be determined by social, political, and psychological factors and boundaries. In fact, studies by Lim (2003) and Smith (1999) show that the boundaries of real knowledge spillovers always match the geographic units. Neighboring geographical units are just partial benefits of knowledge spillovers (regardless of whether it is a country, state, or metropolitan area). As an example, short distances would be of benefit if local knowledge spillovers in Los Angeles could benefit firms in San Diego, which is less than 3 h' drive away. Similar conclusions were reached by Anselin et al. (1997), who recorded the effect of interaction between spatial units and found evidence of knowledge spillovers at both state and MSA levels.

The subject of inter- or extra-regional knowledge spillovers at the European regional level has been relatively poorly investigated. Two studies are worth mentioning here: one by Greunz (2003) and Rodriguez-Pose and Crescenzi (2006). The study by Greunz (2003) aimed to test the importance of geographical and technological proximity for regional knowledge spillovers in EU-15 Member States. The author finds a positive significant effect of innovative efforts pursued in the first- and the second-order neighboring regions on local patenting activity (190 miles or 306 km, on average).

The research by Rodríguez-Pose and Crescenzi (2006) aimed to combine the model of R&D, spillovers and RIS approaches into one, for the EU-25 to highlight the importance of proximity for the transmission of economically productive knowledge. The regression not only confirms that knowledge flowing between neighboring regions improves regional growth rates but also shows that spillovers are geographically bounded. The authors included the educational attainment level of a region (measured by the percentage of the population and of the labor force having completed higher education) and participation in lifelong learning programmes as a measure for the accumulation of skills at the local level. However, the role of lifelong learning, the percentage of labor working in agriculture, the level of long-term unemployment, and the demographic structure of the population turned out to be not significant. In the EU-25 context, only innovative efforts pursued within 180-min' worth of travel time radius have a positive and significant impact on regional growth performance. The regression results confirmed the importance of geographical proximity for knowledge spillovers. Agents within driving distance of one another can exchange information face to face potentially on a daily basis, at a much lower marginal cost than those for whom an overnight stay would be necessary (Sonn and Storper 2005).

The literature has approached knowledge spillovers in the technological space in different ways. Jaffe (1986) can be considered a pioneer in the field of empirical investigation of technological proximity spillovers. He used corporate patent data to compute the distribution of patenting across technology classes in order to evaluate technological similarities between firms. The flow of spillover knowledge to one particular firm is computed as the total of all other firms' R&D expenditures, using the proximity measures as sums. The author found evidence for technology-dependent spillovers controlling industry-specific technological opportunities. Firms active in research-intensive technology groups enjoy higher returns on their R&D expenditures. Jaffe (1986) also employs a clustering procedure to identify groups of firms who presumably face similar technological opportunities. However, as Cohen and Levin (1989, p. 1,084) point out, these cluster variables do not perform better than conventional industry dummy variables.

Several European authors have attempted to proxy technological knowledge spillovers using the KPF framework. Bottazzi and Peri (1999) attempted an empirical investigation in this field and considered a model in the spirit of the endogenous growth literature and allowed for knowledge spillovers across European NUTS I regions. Using the average number of patents over the 1977–1995 period and the average total R&D corporate expenditure for the same period, they focus on the estimation of a spatial diffusion pattern. By dividing space into intervals (from 0 to 400 km) and estimating the impact of each interval on patenting, they find evidence that spillovers of R&D across regions exist, are significant, but decrease rather quickly with distance. Using Jaffe's measure of technological distance between regions and for 30 technological classes, they also test for knowledge spillovers due to technological proximity. According to the authors, technological proximity between regions highly influences patenting activity. However, a limitation of the authors' study is that they assumed accumulation of new knowledge is constant and even over the period considered.

When investigating knowledge spillovers in technological space, Greunz (2003) identifies "technological neighborhoods" using Jaffe's measure of technological distance. It is calculated by means of patent application data to the EPO disaggregated into 118 IPC classes. By dividing space into intervals, she introduces a first, second, third, and fourth-order geographical neighborhood distance matrix (with the first-order neighborhood being an average distance of 109 miles; and the second and third 190 and 274 miles, respectively). She concludes that national border regions (second-order geographical proximity) are extremely important and highly significant for knowledge spillovers. Furthermore, she concludes that technological proximity significantly influences home region patenting activity. This outcome suggests that, on average in the EU, national border regions are relatively far from the country's technological and economic centers. In general, her results indicate that technological proximity between EU regions is at least as important as geographical proximity for the generation of new knowledge.

In summary, the numerous attempts to measure geographical and technological knowledge spillovers at intra- and inter -regional EU levels based on Jaffe's KPF framework provide relative evidence of the importance of proximity for both geographically and technologically mediated knowledge spillovers. However, to the best of the author's knowledge, there hasn't yet been an empirical study that investigates geographical and technological proximity spillovers in the EU-25 regions. Even at the EU-15 level, the scarcity of research efforts in this field has prevented the formation of any commonly accepted conclusion. A systematic comparison of present and past empirical evidence for geographically and technologically mediated knowledge spillovers with the changing EU spatial patterns is important not only for the sake of measuring the progress of integration of the EU research space but also the on-going processes of EU integration and globalization.

4.3 Spatial Knowledge Spillovers and University Research in Empirical Studies Literature

4.3.1 Role of Universities in Knowledge Spillovers in the High-Tech Sector

Considering the importance of universities in spatial and technologically mediated knowledge spillovers, this section discusses the empirical literature's findings on the role local universities play in the dissemination of scientific and technological knowledge via formal R&D cooperation, flow of graduates, and informal social networks. The role of university-industry collaboration networks in geographically mediated knowledge spillovers has been emphasized and evidenced by a number of studies conducted by Anselin et al. (1997), Bania et al. (1993), Baptista (2001), and Ponds et al. (2009). A relatively small number of researchers has investigated the role of technological proximity or technological relatedness in universities' knowledge spillovers; these include the works of Petruzzelli (2009), Brockhoff et al. (1991), Lane and Lubatkin (1998) and Brockhoff and Teichert (1995). Their major findings are discussed below.

The study of Petruzzelli (2009) tests 796 R&D university-industry collaboration cases in 27 universities located in 12 different European countries. His results suggest that innovation value has an inverted, U-shaped relation to partners' technological relatedness, which means that in order to increase innovative performance, a certain threshold of similar technological competencies between partners are required. However, too much similarity may in turn have a detrimental effect on innovative performance since the development of valuable innovations requires dissimilar but complementary sources of knowledge.

Lane and Lubatkin (1998) test pharmaceutical-biotechnology R&D alliances and show that university R&D organizations with greater technological relatedness have a greater innovative capacity, and hence are more likely to observe a higher degree of knowledge spillovers. This is related to technical and market competencies organizations have acquired when dealing with specific technologies and markets. They suggest the presence of overlapping technological skills important for transferring existing knowledge from one organization to another, and for jointly creating new knowledge.

The study by Brockhoff et al. (1991) captures synergistic gains from the exchange of complementary technical knowledge between given bodies and highlights that it is an important driver of R&D collaborations in Germany. Meanwhile, Nooteboom (2000) argues that, on the one hand, too little technological distance may imply a lack of sources of novelties, but on the other hand, too much technological distance may cause problems in terms of communication and mutual understanding between given collaborating institutions. The above-mentioned outcomes suggest there may perhaps be an optimal amount of technology overlap between partners that would affect the potential benefits of collaboration.

Research mobility and social networks can also play an important role in all intra-, inter- and extra-regional knowledge flows. Universities act as 'anchor institutions' by attracting and retaining talented researchers from around the globe. Ponds et al. (2009) used an extended knowledge production function framework and applied it to regions within The Netherlands to carry out their study. The authors found that the impact of academic research on regional innovation is affected not only by geographical proximity but also by social networks stemming from collaboration networks and global R&D networks. Universities can provide local firms with access to the best global research available at any given moment. This has also been discussed by Saxenian (2006), who investigated the process of outsourcing and subcontracting from many developing regions, in the field of R&D. The increasing demand for skilled scientists and rising R&D costs have forced many firms to expand their research networks to link with many developing and transition economies, especially in engineering fields related to electronics and software. Obviously, regions with the largest reserves of scientifically and technically trained manpower, good communication infrastructure, and high knowledge-absorptive goods take better advantage of these global R&D networks.

4.3.2 *Intellectual Property Rights and Knowledge Sharing in the High-Tech Sector*

The willingness of a university to collaborate with industry may also be determined by the intellectual property regime (patent and copyright law). Most research studies carried out in this field have provided some controversial evidence concerning the role of patents and non-patent activities in knowledge diffusion (Baum and Silverman 1998; Kneller 2001; Thumm 2004). A recent analysis of university patenting (Arora and Merges 2004) has been heralded in some literature (Gambardella 2005), where it is suggested that university patenting is not as effective as a mechanism for knowledge diffusion as David (1994) presumed it to be. There are, indeed, as many proponents as there are opponents of patents as a knowledge transfer channel. The proponents of patents emphasize that since the Bayh-Dole Act in 1980, there has been a substantial increase in patents licensing to the industry and economic returns (Kneller 1999, 2001). In this sense, in terms of patents, both the university involved and industry are seen to play active roles in cross-boundary knowledge transfer (Jaffe et al. 1998). In contrast, the opponents reject these claims and argue that knowledge diffusion and its relationship to economic growth cannot be associated with patenting activities (Sargossi and de la Potterie 2003). Exclusive licensing may generate some profits through patents, but is likely to impede future exploration due to the current cost of exploitation (Shulman 1999). Unnecessary and excessive litigation by universities against the industry may prevent the application of new discoveries (Shane 2005). Hence, pure patent-based economic values may endanger the explorative-exploitative imbalance.

The attitude towards patents as a means of knowledge transfer may be determined by the specific technological field, such as biotechnology, semiconductors, aerospace, etc. that we are dealing with. Caulfield et al. (2006) show that about a half of researchers view patents negatively in terms of their impact on the research environment when they are subject to increased secrecy. On the other hand, Walsh et al. (2003) in their study of the impact of patenting of research tools in biomedicine on innovation, found that university research had not been significantly impeded by an increase in patenting; with the exception of patented genetic diagnostics. Relying on the analysis of citation rates of scientific publications before and after the granting of associated patents, Murray and Stern (2007) tested the anti-commons hypothesis and found a modest amount of evidence of the restrictive impact of patents on knowledge diffusion.

Patenting and R&D collaborations alone may not be sufficient to explain knowledge diffusion among researchers in high-tech fields. Several empirical studies have highlighted the importance of cultural proximity and competition-related attitudes as enabling factors of R&D collaboration (Petruzzelli 2009; Hargadoorn 2002; Van Everdingen and Waarts 2003). For example, Petruzzelli 2009 argues that universities and firms belonging to similar cultural groups, and which have had previous ties with each other, are more likely to achieve better innovative outcomes.

Furthermore, a number of scholars have explored ways in which the engagement of researchers' secrecy and publication delays could undermine their contribution to open science and could hamper knowledge-transfer activities (Dasgupta and David 1994; Geuna and Nesta 2006). These studies point to the existence of data and materials that are being withheld among researchers. When it comes to involvement in the commercialization of university-based research, it is shown to be significantly associated with the increased likelihood of data withholding (confirmed by Campbell et al. 2002; Walsh et al. 2007). A study by Davis et al. (2011) confirms the scepticism of life scientists regarding the impact of university patenting on academic research and the norms of open science. Furthermore, in a study aimed at revealing the reasons behind two forms of data withholding, publication delays, and refusals to share biomaterials and data, Blumenthal et al. (1997) found that involvement in academic-industry research relationships and engagement in the commercialization of university research are both associated with publication delays. The latter is associated with the refusal to share research results upon request. A more recent study of geneticists and other life scientists (Blumenthal et al. 2006) shows that not only industry research support and commercialization endeavor, but also other industry involvements, such as consulting or equity, have an adverse effect on verbal data sharing or publishing data in life sciences.

In all, there is little empirical evidence on the role of universities in knowledge spillovers, especially in the context of the role of geographical space in knowledge spillovers in high-tech industries. Most studies examining the diffusion channels of university research knowledge flows focus on only one or a few channels of knowledge diffusion and provide inconclusive results in terms of the impact of university R&D networks on knowledge dissemination or social networking between researchers. Some recent studies admit that empirical findings with reference to knowledge diffusion channels have been poorly represented in the literature and have often produced ambiguous conclusions, for example Walsh et al. (2007), Agrawal and Henderson (2002), Van Looy et al. (2006), Fabrizio and Di Minin (2008). Consequently, there is a need for further research on the nature of knowledge diffusion and transfer activities in the scientific community.

4.4 Evidence of Research Networking in the European Research Area of the EU

Relatively few empirical studies have investigated the EU spatial dimension of R&D networks. This may be because of the informal nature of most of the R&D networks between individual researchers and R&D labs. In recent decades these collaborative networks have become more formalized and therefore it has become easier to keep track of them due to the EU-funded Framework Programs (FP)[1] that have allowed for more detailed analysis of networks' topology, their evolution over time, and their efficiency as mechanisms of knowledge diffusion. Recent contributions in this field include the works of Maggioni et al. (2007), Maggioni and Uberti (2007), Scherngell and Barber (2009), Protogerou et al. (2011), Autant-Bernard et al. (2007), Constantelou et al. (2004), Scherngell and Barber (2009), and Ortega and Aguillo (2010). The below section presents the major findings of some of the above- named titles.

The research project of Maggioni et al. (2007) and Maggioni and Uberti (2007) aims to explain knowledge flows based on the experience of the fifth EU Framework Programme (FP5) research networks. Maggioni and Uberti (2007) modeled

[1] The main goal of the ERA is to strengthen networking between research partners and across research disciplines as well as to link geographically-distant centers of excellence and to disseminate knowledge across Europe. Data on research networks financed by the EU within the different FP are publicly available through the CORDIS.

cross-region collaboration in FP5 programs for five large EU countries, with gravity equations estimated by using standard OLS estimation procedures. The authors concluded that despite the intentions of FP to integrate the EU R&D landscape, the number of links between organizations within the fifth FP funding schemes tends to decay with geographical distance and language barriers. Constantelou et al. (2004), who studied international linkages in EU Framework Programmes, reported similar findings. The results unveiled a picture of significant collaborative activity among clusters of neighboring countries.

The work of Protogerou, Caloghirou, and Siokas (2011) examined the structural characteristics, evolution, and research collaboration patterns in ICT research projects funded by the EU during the sixth and seventh Framework Programmes. The results of this work show clearly that the majority of research networks appear to be quite close together in the network space. Scherngell and Barber (2009, 2010) have studied the importance of spatial and other types of impediments for collaboration within the fifth FPs and concluded that – despite the intended European character – the number of links between organizations in the fifth FP funding scheme tends to decay with geographical distance and language barriers.

The study of Scherngell and Barber (2009) contributed to the existing empirical literature, by investigating cross-region R&D collaboration networks in Europe within FP5 by adopting a Poisson spatial interaction modelling perspective. The research included the 25 pre-2007 EU Member States (without Malta and Cyprus). Their model results show that geographical and technological distances between two R&D organizations have a significant negative effect on the likelihood of their mutual collaboration. This implies that it is most likely that cross-region R&D collaborations occur between regions that are close to each other in terms of technology. This finding is somewhat similar to the results obtained by LeSage et al. (2007), who studied the importance of interregional knowledge spillovers. The only significant difference is that technological distance effect of the latter authors is much higher for interregional FP collaborations than in the study of Scherngell and Barber (2009). Furthermore, the outcome of LeSage et al. (2007) shows that for collaborations amongst public research organizations, negative effects of geography nearly vanish.

Scherngell and Lata (2011) monitored progress towards ERA by identifying collaboration intensities between 255 European regions participating in FP over the period 1999–2006. The authors employ spatial interaction models accounting for spatial autocorrelation by using spatial filtering methods. They concluded that the FPs help to increase the probability of long- distance R&D collaborations and contribute to geographically integrated European research systems. Taking into account the "cultural and language barriers," the authors obtained some mixed results. In general, language barriers seem to be fading; however, this is happening relatively slowly. Similarly, Roediger-Schluga and Barber (2006) and Kruckenberg et al. (2008) found that the integration between already collaborating organizations increased over time by using social network analysis techniques and concluded that these findings point to a trend towards a more integrated ERA.

Other interesting findings are presented in the studies of Autant-Bernard et al. (2007) and Breschi and Cusmano (2004). Autant-Bernard et al. (2007), who compare geographical and social distance effects on collaborative patterns within the 6FP in micro- and nanotechnology industries in the French R&D landscape. Their results show that geographical distance is not the main determinant of collaboration at the European level. Instead, they claim that social distance (previous R&D collaboration and friendships) matters for sustaining R&D networks. On the other hand, Breschi and Cusmano (2004) find that many participants of the current FP projects also participated in previous FPs. Based on this finding the authors conclude that it can be rather difficult for new participants (from new EU Member States) to join the FP network. Olechnicka also concluded that researchers from the EU regions of strong scientific performance cooperate with equally strong regional partners, and not with those of the weaker regions, because this participation in the R&D network requires an already existing critical mass of scientific activity effects. For these reasons, attempts to participate in FP projects more often fail in the case of new Member States compared to the already experienced R&D units of EU-15 Member States (2012). Generally, these empirical studies show that technological distances between two R&D organizations have a negative impact on the likelihood that they will collaborate. In terms of geographical proximity, the majority of studies show that R&D networks usually tend to be localized and therefore confirm the significance of geographical distance for cross-region R&D networks. Their results suggest that different kinds of geographical space, such as physical distance between regions and the existence of country borders between regions or neighboring regions affects R&D cooperation in Europe. A relatively small number of studies have pointed to a trend towards a more integrated ERA. Finally, another group of studies emphasizes the importance of social networking and previously established R&D friendships as two of the major reasons for current R&D collaborations. Hence, overall, the results of all existing studies and research on the role of geographical and technological proximity in EU FP are inconclusive and fragmented, with reference to the experience of selected scientific fields. The current study attempts to fill, at least partially, the gap in the existing literature by investigating the role of geographically and technologically mediated knowledge spillovers for patenting in the high-tech industry. This is done firstly by applying a spatial econometric model for the EU regions during 1998–2008, then by presenting the findings of the case study in interviews with academic scientists and engineers from the six high-tech industries (Aerospace, Biotechnology, Communication, Computers, Semiconductors, and Lasers).

4.5 Knowledge Spillovers in the EU Regions – Evidence from the High-Tech Sector

4.5.1 Selected Methods of Analysis

In the above sections the author discussed different measures of knowledge spillovers in terms of the accessibility of regions. In order to incorporate geographical accessibility for the purpose of the current study, the actual travel times between locations are used to discount spatial knowledge spillovers. To clarify, geographical proximity is measured in kilometres using actual travel time between locations (Beckmann 2000). The reason for applying the traveling time distances method is because it shows rather realistic estimates of the possibilities of physical access (and costs), rather than the distance itself. It also reflects major differences in regional communication and road infrastructure. For example, a long travel time may indicate that the average speed in the regional road network is low, which suggests that the increase in travel time due to a link closure would be significant, even if the increase in travel distance is only moderate (Jenelius 2010). The inability to reveal such disparities is a major drawback of conventional geographical distance. Time distances are also crucial when it comes to attending to business meetings and to determine the spatial borders of labor markets (Johansson and Klaesson 2001).

Geographical knowledge spillover efforts are weighted by a row standardised square inverse distance weight matrix, reflecting that the intensity of interactions declines with increasing distance. More precisely, the weight[2] attributed to the R&D expenditure of region j when investigating patenting activity of region i, is given by:

$$f(c_{ij}) = w_{ij} = \frac{\frac{1}{d_{ij}}}{\sum_{j} \frac{1}{d_{ij}}}$$

where:

i– home region; j- neighboring region.

Distance d is calculated (is the effort, cost/opportunity) as the average trip length (in minutes) between region i and j, taking regional capitals as regional references. The calculation of distance is based on the analysis of travel time. The road distance is used, rather than the distance from A to B measured in a straight line, as it gives a more realistic representation (particularly on a smaller scale) of the real 'cost' of interaction and contacts across space. The author approaches geographical neighborhood by an "nth-order" continuity matrix.

The first-order continuity matrix takes into consideration the direct neighbors of a given region X, whereas the second-order continuity matrix accounts for the

[2] The knowledge spillovers matrix was based on Greunz (2003).

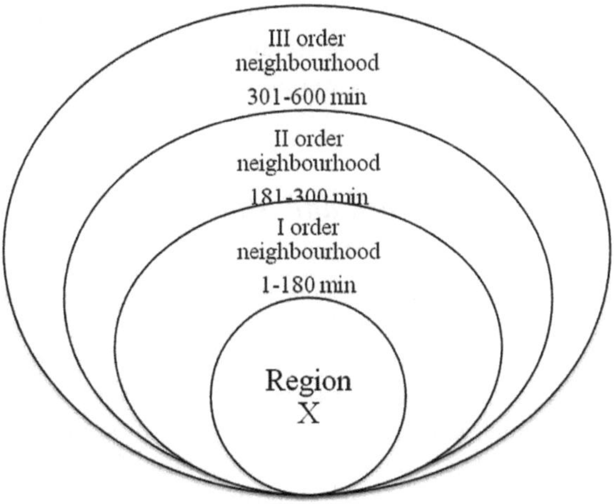

Fig. 4.1 Geographical proximity of region X (Source: own illustration)

neighbors of the neighbors, as well, and so on (Fig. 4.1). The first matrix considers a region to be a neighbor if the travel time there is less than 180 min (W0–180 min – first-order neighborhood), the second one considers the range between 180 and 300 (W180–300 min – second-order neighborhood) and in the third one a region is considered to be a neighbor if it is located in a range between 300 and 600 (W300–600 – third-order neighborhood).

The "technological neighborhood," as applied after Jaffe's measure of technological distance, is calculated by means of high-tech patent application data to the EPO disaggregated into six industries (according to IPC classes). This measure, which is also used to weigh R&D expenditures of "technological neighborhood," is defined in the following way:

$$p_{ij} = \frac{\sum_{k=1} f_{ik}^{G} f_{jk}}{\sqrt{\left(\sum_{k=1}^{G} f_{ik}^{2} \sum_{k=1}^{G} f_{jk}^{2}\right)}}$$

Where: six high-tech industries; *i*– region; *j*- sector;

fik is the share of a particular high-tech patent class *k* in the total of high-tech patents of region *i*. If the technological profiles of region *i* and *j* are similar, *pij* is close to one. Conversely, the more regions are technologically different, the closer *pij* is to zero. The first-order technological neighborhood of region *i* is built up of three regions that display the highest values of *pij* independent of their geographical location. The second-order technological neighbors of the concerned region *i* are the three regions, which show the second-highest values for *pij* (and so on). Similarly, as in the case of geographical proximity, the cut-off order of technological neighborhood is determined by estimation. When taking into account the geographical distance that separates region i and its technological neighbors, the R&D expenditures of technological neighbors have also been weighted by the square inverse distance weight matrix previously defined.

4.5.2 Framework and Model

The adopted modelling approach used to assess intra-, inter-, and extra-regional knowledge spillovers is derived from KPF by Griliches (1979), with its later modifications that also accommodate the spatial dimension (Jaffe 1989; Feldman 1994).

The R&D output is a proxy for high-tech patent applications per capita to the EPO, which is the only available innovation measure at the EU NUTS2 regional level. Patent data refer to the living place of the inventor. R&D input is approached by R&D expenditures per capita expressed in PPS per capita at constant 2,000 prices.

The aim of the modelling is to answer to what extent patent activity of home regions is determined by the home and neighboring regions' R&D expenditure (intra-, inter-, extra- R&D flows). In order to test for knowledge spillovers generated by the geographical and technological neighborhood, the author tested the below specification, which is based on the relation summarized in Eqs. 4.1 and 4.2:

$$\begin{aligned}\ln(P_t) = c + \alpha(\ln(R\&D_t)w_{ij}) + \beta\left(\sum_{\substack{gn1=1\\ gn1\neq j}}^{GN1} \ln(R\&D_{gn1})w_{t,gn1}\right) + \ldots \\ + \delta\left(\sum_{\substack{gnj=1\\ gnj\neq j}}^{GNJ} \ln(R\&D_{gnj})w_{t,gnj}\right) \\ + \ldots + \varphi\left(\sum_{\substack{gnj=1\\ gnj\neq t}}^{GNJ} \ln(R\&D_{gnj})w_{t,gnj}\right) + \psi \ln \mathrm{HKIS} + \chi \ln \mathrm{HQ} + t_i\end{aligned} \tag{4.1}$$

$$\ln(P_t) = c + \alpha(\ln(R\,\&\,D_t)) + \kappa\left(\sum_{\substack{tn1=1\\tn1\neq t}}^{TN1} \ln(R\&D_{tn1})p_{t,tn1}\right) + \dots$$

$$+\,\nu\left(\sum_{\substack{tnj=1\\tnj\neq t}}^{TNJ} \ln(R\&D_{tnj})p_{t,tnj}\right)$$

$$+\dots+\tau\left(\sum_{\substack{tnj=1\\tnj\neq t}}^{TNJ} \ln(R\&D_{tnj})p_{t,tnj}\right) + \psi\ \ln\ \mathrm{HKIS} + \chi\ \ln\ \mathrm{HQ} + t_i \qquad (4.2)$$

Where:

P is the number of high-tech patents per 1,000 inhabitants;
$i = 1\dots.N$ indexes the spatial units of observation, $i = 1\dots.186$;
$gnj = 1\dots GNj$ indexes the geographical neighbors of region $i, j = 1\dots.$J;
$tnj = 1\dots TNj$ indexes technologically similar regions with respect to the technological profile of region $i, j = 1\ \dots\ J$
w is a row standardized square inverse distance weight matrix between regions i and gnj with elements $w_{ij} = \frac{\frac{1}{d_{ij}}}{\sum_j \frac{1}{d_{ij}}}$ where the distance d is calculated using spherical trigonometry and taking regional capitals as regional references.

R&D expenditures of technological neighbors are weighted by *pi, tnj*, that is, Jaffe's measure of technological distance applied to six high-tech classes over the 1998–2007 period. According to this measure, different order neighborhood groups must be determined.

Other variables:

HKIS – High-tech knowledge-intensive services as a share of total services (1999–2008);
HQ – students of tertiary education at ISCED levels 5–6-as a percentage of all pupils and students at the regional level (1997–2007);

Following Jaffe (1986), the potential interaction between university and corporate R&D departments is captured by extending the base Eq. 4.1 in Chap. 1 to capture the effects of R&D expenditures, the number of students at the second stage of tertiary education (HQ) and the share of high-tech knowledge-intensive services (HKIS) on high-tech patents. The number of HQ proxy, the size of the academic sector, or the role of universities in LKS and knowledge-absorption and transfer capacities of regions. The presence of an HKIS as a proxy for a region's knowledge absorption and diffusion capacities in the high-tech industries (including the potential effect of outsourcing).

The model is tested for the 1998–2007 period, on the sample of 186 EU regions investigated in Chap. 3. It should be noted that while the majority of regions selected for the research sample is on the NUTS II level, as this was the lowest administrative unit for which the necessary data was available for the selected period of research. Econometric estimations are performed by means of a fixed effect panel data estimation method.

4.5.3 Estimates Results

In general, the model outcome suggests that "home" R&D is highly significant for regional high-tech patent activity. This result in Table 4.1 was able to confirm the dominant literature, which suggests geographical clustering matters for successful collaboration for innovation, and more generally for knowledge transfer and LKS. Moreover, the model outcome supports the idea that high-tech innovations also require integrating inter- and extra-regional R&D efforts. This is in accordance with past studies' results on a broader group of patents, which proved that geographical and technological neighborhoods' R&D expenditures (inter-regional knowledge flows) strongly influence the number of patents in the home region Rodriguez-Pose and Crescenzi (2006), Greunz (2003), and Bottazzi and Peri (1999).

The fact that patenting activity (P) turned out to be a significant negative function of knowledge spillovers from its second-order geographical neighbors may indicate little synergistic gains in exchanging knowledge between the neighboring RISs. The latter may result from either too much overlapping of the bodies' technological skills or a too high technological difference between them, which causes problems in communication or mutual understanding. This in turn suggests that there is no optimal amount of technology overlap between partners that affects both partners' potential benefits from collaboration. On the other hand, a positive function of knowledge spillovers from its third-order geographical neighbors may imply the increasing role of more long-distance knowledge flows for high-tech innovations. The latter may result from the existence of R&D collaboration networks within FPs and among scientists.

As explained in the previous sections, the ability of a region to make productive use of another region's knowledge depends on the degree of technological similarity between their productive systems of knowledge-intensive activity. The regression was also considered to measure how the distance factor affects knowledge spillovers between technological neighbors. In reference to technologically mediated knowledge spillovers, the outcome of the regression suggests that when purely technological externalities are weighted by the geographical distance, the third-order geographical proximity (extra-regional knowledge flows) turns out to be extremely important and significant for regions' patenting activity in the high-tech sector.

Table 4.1 Model with spillover matrix: estimates of knowledge spillovers in the geographical and technological spaces

	Geographic spillovers		Technological spillovers		Geographical and technological spillovers	
Variables	Coeff.	T-stat.	Coeff.	T-stat.	Coeff.	T-stat.
Logrdhome_region	1.192	(17.07)***	1.185	(17.12)***	1.191	(16.81)***
Loghq	−0.112	(−0.95)	−0.141	(−1.23)	−0.128	(−1.11)
Loghkis	0.462	(3.90)***	0.396	(3.49)***	0.423	(3.55)***
Logrdgeospillover1	−0.013	(−0.72)			−0.013	(−0.71)
Logrdgeospillover2	−0.029	(−2.53)**			−0.024	(−2.02)**
Logrdgeospillover3	0.016	(1.98)**			0.015	(1.80)*
Logrdtechspillover1			0.001	(0.10)	0.000	(0.02)
Logrdtechspillover2			−0.001	(−0.21)	−0.000	(−0.07)
Logrdtechspillover3			0.013	(2.53)**	0.012	(2.27)**
_cons	−4.985	(−12.58)***	−4.924	(−12.44)***	−4.967	(−12.57)***
N	649.000		649.000		649.000	
r2	0.610		0.611		0.613	

Source: own estimations
Standard errors in parentheses: ***$p < 0.01$, **$p < 0.05$, *$p < 0.1$

These findings may suggest that knowledge spillovers are greatest among regions that specialize in similar technology areas, regardless of geographic proximity. This implies that European R&D networks may be of greater importance as sources of knowledge than inter-regional knowledge flows for the innovative activities of home regions.

In addition, the presence of HKIS is highly significant for the generation of new knowledge in home regions and confirms the importance of absorption and transfer capacities of knowledge in high-tech industries. The importance of HKIS for local innovative activity may also suggest partial outsourcing of R&D and high skills to other regions.

The estimated results of the role of the academic sector (share of students of tertiary education programs) in high-tech patenting turned out to be insignificant. This outcome is quite surprising and is not in line with the evidence provided by a number of studies (Anselin et al. 1997; Bania et al. 1993; Baptista 2001; Ponds et al. 2009). The results may suggest that the academic sector doesn't play an active role in high-tech patenting, or university faculties have a negative attitude towards patents as major knowledge diffusion channels (Baum and Silverman 1998; Kneller 2001; Thumm 2004; Shulman 1999). The results may originate from flaws in the model and measures: regional boundaries may not be a poor proxy of NUTS2 as some consider areas such as NUTS1 to be too large. Also, the relationship between the academic sector and patenting would have been stronger if only technical university students had been considered. Finally, patents may not capture the real innovative output of R&D efforts, as many innovations in high-tech industries are not being patented. The results would have been different if less aggregated technological areas were considered.

The above tested model can only capture the general trend for the high-tech sector and doesn't suggest whether geographical and technological proximity affects innovation processes in specific high-tech industries or all the localities. It is important to provide better evidence of whether the determinants of proximity effects are specific to the particular scientific field. The latter couldn't be achieved without using qualitative methods of research. The next section will complement the estimates outcomes with the interviews-based research output, and provide a more detailed analysis of the patterns of geographically and technological mediated knowledge spillovers in the high-tech industries.

4.6 University-Based Knowledge Spillovers in High-Tech Fields. The Interview Results

To understand the complex system of knowledge spillovers in high-tech clusters and to highlight the role of geographical and technological proximities for the EU-10 regions, the model output was extended and strengthened with 35 in-depth interviews with heads of R&D units (usually doctors or professors) in one of the six high-tech industries (Aerospace, Biotechnology, Communication, Computers, Semiconductors, and Medical Devices and Lasers). However, qualitative interviewing as a method of research has some downsides, mainly because of the general belief that the study of a limited number of cases can offer no solid ground and cannot be deemed a reliable method to establish generality of findings. Nevertheless, in such complex and inter-disciplinary fields as the high-tech industry there seems to be no better method that could be applied. Interviews were conducted between September and October 2011.

The purpose of the interviews was to determine the role of geographical space in knowledge diffusion, spillovers, and R&D networks. To what extent are engineers in high-technology disciplines sourcing and diffusing knowledge through seminar and conference events, collaboration with the business sector, or in social networks with researchers? What is the attitude of researchers towards open science and sharing research results?

The author conducted interviews with representatives from the following universities[3]: University of Liege (Région Wallonne), Universite Libre de Bruxelles (Region Bruxelles) Department of Aerospace Engineering of Czech Technical University (Prague), Rzeszów University of Technology (Karpackie region), Budapest University of Technology (Közép-Magyarország), Flanders Institute for Biotechnology (Flanders region), GIGA Institute of Biology (Région wallonne), Wrocław Medical University (Dolnośląskie region), Eötvös Lóránd

[3] The respondents were approached by a personal contact, email, and a phone call.

University (Közép-Magyarország region), Poznań University of Life Science (Lubelskie region), Faculty of Molecular Biology of Warsaw University (Mazowieckie region), Zwijnaarde Science Park (Flanders region), Department of Computer Studies of Tallin University of Technology (Estonia), Biotechnology Innovation Institute/Cluster (Dél - Dunántúl), Gdańsk Technical University (Pomorskie region), Department of Computer Technologies of Czech Technical University (Prague region), Grenoble Institute of Technology (Rhône-Alpes region), Łódź Technology University (Łódzkie region), Prof. Dr. Ferenc Krausz (Max Planck-Institute of Quantum Optics), Prof. Riku Jäntti (School of Electrical Engineering, Aalto University), Prof. Peter Göhner (Institute of Industrial Automation and Software Engineering, University of Stuttgart).

The criteria for the selection of the case studies that were to be used for the research project were mainly geographic as that would allow for a more in-depth analysis of the EU-10 experience, since this group of regions has the biggest shortages in the empirical literature. The field interviews were carried out in different EU regions with a set of carefully selected R&D organizations, according to their highest performance in their relative scientific fields. Although the case study selection attempted to consider an example from each Member State, some regions turned out to be represented more extensively, because of their higher patent performance in several high-tech disciplines. Since the number and degree of development of the EU-15 clusters is much more advanced, this number is more favorable for the EU-15 regions. The brief results of the cross-sectional survey study of the EU high-tech university-based knowledge flows is presented in Annex 3.

The survey questionnaire consisted of five sections denoted as A, B, C, D, and E. The first section (A) contains general information about the respondents (names, scientific title), their S&T field, as well as their major source of R&D funding. The second section (B) aims to capture the main channels of knowledge diffusion (seminars, conferences, workshops, collaboration with the business sector, publishing activity, patents, and informal (face-to-face) contact). The third section (C) covers questions related to geographically and technologically mediated knowledge spillovers, such as the significance of intra- (local), inter-, extra-(European) and global sources of knowledge, and asks about the type of knowledge: technological, market related, managerial, and other. Technological knowledge refers to all kinds of technical information, specifications, and know-how necessary to create and produce a product. Innovation and the exchange of technological knowledge are of utmost importance. Market knowledge refers to information on future market developments, potential customers, and demand, which is crucial for firms to create and sell their products.

The fourth section (D) assesses the participation of the respondents in an R&D unit/lab in knowledge and technological networks within the home institution and in home region institutions, other institutions within its national boundaries, other institutions in the EU, and in other countries. Finally, the fifth section (E) treats issues related to secrecy in university research, such as access and sharing research results or materials (software, genetic sequences, data) with other scientists. A brief overview of the survey results for each technological field is presented below.

4.6.1 (A) and (B) R&D Funding and Major Channels of Knowledge Diffusion

The research activity of the selected R&D institutions included the space and aviation technology fields. Although the two fields are interrelated, there are significant differences between their knowledge bases. Space engineering relies more on an analytical (scientific) knowledge base, while aviation (aeronautics) depends more on synthetic knowledge. Public-funded (or EU) R&D is dominant in the first instance and private and university R&D predominate in the second.

In the view of field experts, for space engineering R&D activity, the major knowledge diffusion channel was through publishing and informal (face-to-face) contacts. The latter is especially important since space engineering and R&D-related activities source their knowledge from other related industries such as IT, chemistry, etc.; therefore, there is a need for experienced scientists in the home region of the relevant industry. On the other hand, in the case of R&D activities in the field of aviation, which is characterized by the synthetic nature of its knowledge base, the experts mentioned collaboration with business, along with workshop-related activities, as the major forms of knowledge diffusion.

The major source of financing of biotechnology-related R&D activities was public funds, thus it is not surprising that the major portion of biotechnology research is conducted mainly in the university labs. Industry financing of R&D in biotechnology labs was dominant only in the case of respondents from biotech labs located in scientific parks or clusters in the Flanders region (Zwijnaarde Science Park) and the Transdanubian region (Biotechnology Innovation Cluster). Publishing activity, patents, and collaboration with the business sector were the most common channels of knowledge diffusion in this industry in their respective order. Furthermore, technological convergence in the biotechnology-pharmaceutical industry and the fact that the biotechnology industry relies very much on basic research, signify the growing importance of university-business R&D partnerships as knowledge-diffusion channels. Such observations had already been made by previous studies conducted by Zucker and Darby (1995) and Audretsch and Stephan (1996). Briefly, they argue that pharmaceutical firms that performed basic research in close co-operation with academia produced more patents. Social networking and informal contacts seem to be a more important process at the beginning of R&D collaboration, as they allow for building credibility between potential partners, but do not always lead to common research projects. Research mobility was another significant channel of knowledge dissemination.

All the surveyed R&D units in the computer and communication industries have had their R&D activity financed by the public, the EU (in the telecommunication field), and private business/industry. In terms of major forms of knowledge diffusion in the field of communication, collaboration with the business sector (both formal and informal) and publishing activity, respectively, were the most important ones. Because the communication field is not a single technological domain but an aggregation of electronics, telecommunications, computer hardware and software, Internet-based contents, and applications and services, innovation may stem largely from the combination or integration of different pieces of knowledge residing in various sources spaced out across sectors and industry segments.

According to the respondents, social networks and informal contacts between engineers in private firms and university researchers serve as channels for sharing knowledge about market characteristics and innovation opportunities. Many such meetings serve as platforms to exchange views and opinions regarding major developments taking place in industry and the Internet, the role of technology and the Internet in society, major challenges, etc.

Finally, R&D financing is one of the biggest challenges in the semiconductor and laser technology industries, since activities within these industries require huge capital investments. In the case of the semiconductor and laser technologies industries, the public sector was responsible for the bulk of R&D funding for the selected R&D units, followed by industry, private R&D, and EU funds. Publishing, collaboration with the business sector, and informal face-to-face contacts were the most common modes of knowledge diffusion in these industries, in this order of importance. Geographical proximity and face-to-face contact regarding technological knowledge exchange provide a great deal of opportunity for R&D cooperation. Some respondents in the laser and medical device fields mentioned that their R&D institutions have formal or informal contacts with local hospitals and companies.

4.6.2 (C) Geographically and Technologically Mediated Knowledge Spillovers

All interviewed scientists considered local knowledge to be significant. For space-related research, both local and global sources of technological and scientific knowledge were of great importance, whereas for the aviation industry, local and inter-regional markets and managerial knowledge played an important role, especially in interactions with clients and suppliers, which is rather typical for industries that are based on synthetic knowledge and knowledge resulting from the learning by doing, using, and interacting. In collaborative relationships with industry, the R&D units interviewed exchanged technological knowledge, together with market and managerial knowledge. The latter could be explained by the complex nature of the innovation processes in the aviation industries and their stronger relationship with the business sector, which requires access to diverse knowledge. Broekel and Boschma (2010) also observed this for the South Holland aviation industry.

Space-related research activity links between universities and industry are important, and occurred more in the field of applied R&D rather than in basic research. These findings could be compared to the results of a similar survey based on studies and an interview conducted by Draganinska and Wink (2005) for the Hamburg region (Hamburg Technical University), Broekel and Boschma (2010) for South Holland (Technical University of Delft), and Cooke and Ehret (2009) for the Wales Region (Cardiff University). Broadly, the findings suggest that the major gap between the intensities of R&D collaborations with industry in the aviation and space engineering fields can be explained by the importance of market knowledge for aviation-based R&D and technological knowledge for space engineering.

For the biotechnology R&D units, local technological knowledge was found to be a major source of knowledge spillovers. Additionally, market and managerial types of knowledge turned out to be equally important for those R&D labs located near biotech clusters. The broad observations described were also confirmed by other surveys such as those by Tödtling and Trippl (2005) for the Vienna region, Häussler and Zademach (2006) for the Munich region, and van Egeraat and Declan Curran (2010) for the Dublin region. Extra-regional (European) and global knowledge flows were the second most important source of information for technological knowledge.

As for communication and computer-related R&D activities, local technological and scientific knowledge flows were important for all the respondents interviewed. For all the R&D units interviewed, local companies and other home R&D institutions were an important source of knowledge sourcing, with the exception of the Tallinn University of Technology (TUT) (as explained below). This is especially true for the tacit (informal) nature of technical and scientific knowledge. The latter implies that the geographical clustering of the computer industry appears to have strengthened rather than weakened over time. These findings are similar to those presented by Peri (2005) and Griffith et al. (2011), who found that knowledge spreads much more quickly across regional boundaries in the computer and communication sector. As for the case of TUT, inter-regional knowledge flows were equally important for knowledge sourcing. TUT carries out contract research for large multinationals such as Nokia and small companies such as Fincitec in Finland that provide the university with relevant technological knowledge. However, most of these R&D-related activities for the Scandinavian ICT contractors were low value added.

Furthermore, local sourcing of market and managerial knowledge does not play a significant role in the process of new knowledge creation in the considered R&D institutions in the field of communication. This was also observed by Ostergaard (2008) for the EU-15 regions. On the contrary, in the case of computer science R&D activities, the experts interviewed considered that market and managerial knowledge had a more local character, while technological knowledge came from elsewhere (inter- or extra-regional sources). One way to explain this is that the computer industry's innovation activities rely more on synthetic (based on customers' views) rather than analytic knowledge (research); therefore, the research-intensive environment may not be relevant for this field. This was previously observed in a Cambridge University-based ICT cluster (East of England) by Stam (2009), as well as in other EU-15 clusters by Giuliani (2007), Morrison and Rabellotti (2009), and Østergaard (2009).

In the opinion of the field experts, technical knowledge in the semiconductor clusters is highly localized, mostly around the universities, whereas market and managerial knowledge has a more global character. This is because the semiconductor industry requires highly technical knowledge covering a range of disciplines, including physics, chemistry, materials science, etc.

Furthermore, according to the respondents, local companies are an important source of new knowledge. This is because the process of designing new semiconductor chips is extremely complex and requires feedback and professional judgment from companies. The development of semiconductor technologies depends upon complex analytics and embedded processes, routines, and knowledge. This is in line with earlier studies, in particular those by Almeida and Kogut (1999), who found that knowledge is highly localized within each semiconductor cluster and seldom spills across locations.

Meanwhile, the interviews showed that the extra-regional (European) and global knowledge flows were a major source of knowledge for the R&D units focusing on laser technologies. The process of knowledge sharing and existing best practices play an important role in R&D collaborations and knowledge networking. This is because the rapid development of the science-based field of laser technology and its applications has an impact on a vast number of industries worldwide and on medicine in particular.[4] The importance of global knowledge sourcing for R&D units in the laser technology industry was emphasized in the works of Bertolotti (2005), Bromberg (1991), and Grupp (2000). Furthermore, university R&D efforts in this field would complement the knowledge of upstream producers, especially in specific laser applications. In the laser systems industry, this is expected to be the case for medical applications and, to a lesser degree, for materials processing.

4.6.3 (D) and (E) R&D Networks and Knowledge Sharing

Since the demand for R&D services in space engineering comes primarily from public bodies, one could expect a higher density of public R&D networks on the national and EU levels in this R&D field. In fact, the respondents mentioned that the presence of other national and EU- based R&D institutions is more common in R&D and knowledge networks in space engineering than in the aviation industry. Most of the EU-10-based R&D units have only recently started to participate in EU-funded R&D projects, resulting from 6 to 7 FP (2002–2006 and 2003–2007, respectively). Examples of these are the AERA-Pro Project, ECARE and ECARE+. Similarly, in terms of R&D collaborations and knowledge networking in the biotechnology and communications fields, home region and nationally based

[4] Bertolotti (2005) refers to lasers as one of the most important scientific inventions of the twentieth century, with a great variety of applications that include range finding and transmission and storage of information, material processing, printing, and medical technology.

R&D institutions and business entities seem to play a more significant role than EU institutions. The latter confirms the general features of a strong spatial concentration of the biotechnology industry. On the other hand, EU ERA-NET-based R&D funding played only a secondary role. In contrast, R&D units conducting research in the computer science-related field considered other EU R&D institutions to be major partners in R&D collaboration and networking.

While the home university is clearly an important source of knowledge spillovers in the semiconductor field, collaboratively aligning with other universities in the country and research networks with the EU and other countries such as the US and Japan was also critical for almost all interviewed experts. In terms of knowledge networking all the field experts emphasized their local character (except for Estonia), where the home institution and other R&D institutions were to be the least important ones. In fact, for Tallinn University, research linkages with neighboring regions were the most important factors.

Finally, the huge capital investments necessary for laser technologies' R&D activities make European-funded projects (ERA-SPOT (Strengthening Photonics and Optical Technology for Europe)) among the major technological platforms and mechanisms for knowledge networking in this field. It is also in the field of laser technologies and biotechnology where field experts have emphasized the importance of IP protection and secrecy in R&D projects and suggested that all the results are to be published, but not however discussed in public or in an informal way.

Summing up, the empirical evidence based on field experts' opinions points us to the conclusion that there is a strong tendency to cluster high-tech R&D activity. However local knowledge sourcing seems to be determined by the type of knowledge itself and may vary from one scientific field to another. This tendency is especially true for R&D fields based on synthetic knowledge as they source their market and managerial type of knowledge locally, whereas for R&D fields based on scientific (analytical) knowledge this is true for the technological type of knowledge. Furthermore, local knowledge sourcing is particularly important in scientific fields relying on R&D-related activities in other related industries, such as IT and chemistry for space engineering research; electronics, telecommunications, computer hardware and software for communication research; biochemistry, pharmacology, agriculture, and IT for biotechnology research. The need for sharing the experience of scientists in these multiple scientific areas partially explains the importance of formal and informal (face-to-face) contacts in the knowledge networks of the home region.

This is not however universally true, since in the case of laser science, technological knowledge sourcing was embedded mostly in the international knowledge networks. One could generalize that for technological advancement in the home region not only local, national or the EU, but knowledge linkages at multiple spatial scales are important and appear to be simultaneous.

However, the ability to build knowledge networks with other national and international partner institutions depends largely on the home intuitions' reliance on their own technological knowledge and the level of existing tacit or high absorptive capacity. The level of internalization of R&D activity and the mobility of researchers seem to be higher in countries with high research capacity, located in EU-15 Member States. In contrast, the respondents from EU-10 Member States with lower research capacities tend to have a lower level of participation in R&D networks and mobility flows. Furthermore, the experts' response suggests that even though proximity facilitated interaction and cooperation through providing a mechanism for coordination, it cannot be synonymous with co-location. The fact that people from the same region share the same language and the same social culture and beliefs may also be of great importance.

While these results may be accused of generalization and of being limited by the boundaries of the researchers' personal experience, they illustrate the critical importance of technological and geographical proximity for universities and any R&D units' research activities and thus reflect the nature of the knowledge flows within the high-tech sector in EU regions.

4.7 Summary and Conclusions

In general, the model outcome suggests that "home" R&D is highly significant for the high-tech patent activity of regions. This result could confirm the dominant literature's finding that geographical clustering matters for successful collaboration for innovation, and more generally for knowledge transfer and LKS. Moreover, the model outcome supports the idea that high-tech innovations also require integrating inter- and extra-regional R&D efforts. This contrasts with past studies on a more broader group of patents, which found that a geographical neighborhood's R&D expenditures (inter-regional knowledge flows) strongly influence the number of patents in the home region Rodriguez-Pose and Crescenzi (2006), Greunz (2003), and Bottazzi and Peri (1999).

The fact that patenting activity turned out to be a negative function of knowledge spillovers from geographical neighbors may indicate either too much overlapping of technological skills or no optimal amount of technology overlap between partners, which affects the potential benefits of collaboration. On the other hand, a positive function of the knowledge spillovers from extra-regional knowledge flows may imply the increasing role of long-distance knowledge networks among scientists. These findings may suggest knowledge spillovers are greatest among regions that specialize in similar technology areas, regardless of geographic proximity. In addition, the presence of HKIS is highly significant for the generation of new knowledge in home regions, and may also suggest partial outsourcing of R&D and high skills to other EU regions.

The empirical evidence based on field experts' opinions indicates that local knowledge sourcing is determined by the type of knowledge itself and may vary from one scientific field to another. This tendency is especially true for R&D fields based on synthetic knowledge from the same or other relevant scientific disciplines (e.g. aviation) as they source their market and managerial type of knowledge locally. Whereas for R&D fields based on a scientific knowledge base (e.g. space engineering) this is true for the technological type of knowledge.

The research findings show that advancement in complex R&D activities in many high-tech fields depends upon sharing the experience of scientists and industry representatives on multiple spatial scales (both formally and informally). Yet, the actual ability of regional R&D units to participate and take advantage of knowledge networks (locally or/and globally), depends largely on their own stock of knowledge and absorptive capacities.

Chapter 5
Localized Knowledge Spillovers, Agglomeration Externalities, and Technological Dynamics in High-Tech Industries. Evidence Based on the EU Regions

5.1 Introduction

This chapter raises the question of whether or not and to what extent the structure of industrial activity generates externalities that influence a given region's innovative activity in high-tech industries. The previous chapter has shown that for the high-tech sector, a home region's knowledge sourcing is very important along with university-industry collaboration. Apart from the universities' role in R&D and knowledge diffusion, the home region's productive system has a crucial influence on regional innovative potential as a whole. Furthermore, firms and R&D institutions are not isolated in their innovation activities but are highly dependent on their external business environment at the sector and regional levels. Therefore, the following chapter tries to take into account the overall productive structure and dynamic externalities in order to analyze a region's potential for innovation activity and technological change.

Several studies have emphasized the role of an underlying productive structure within a geographic unit in its innovative performance. Most of these studies revolve around two key structural elements of productive systems: the degree of diversity (Jacobs externalities) and specialization (MAR externalities). The third key element of any regional productive system is the degree of local competition that may influence the region's knowledge creation capacity (Porter externalities)

Moreover, from a dynamic point of view the role of the regional productive structure in terms of generating technological externalities may differ when a particular technological sector is considered. To the best of the author's knowledge there hasn't been any study to date that considers the role of dynamic externalities and technology maturity in high-tech industries in a comparative way.

By investigating the industrial structure of the EU regions, the following chapter aims to identify the role of local knowledge spillovers and agglomeration externalities in innovation activities in high-tech industries. In addition, the role of industry maturity is discussed for a better understanding of the role of agglomeration externalities in regional innovations and growth.

M. Runiewicz-Wardyn, *Knowledge Flows, Technological Change and Regional Growth in the European Union*, Contributions to Economics,
DOI 10.1007/978-3-319-00342-9_5,

The rest of the chapter is organized as follows. The first section presents an overview of the subject-related literature and defines the different measures that aim to characterize the industrial structure of European regions. The second section of this chapter explains the model and its results for a particular high-tech industry. The conclusion summarizes the most significant findings and suggests future research topics.

5.2 Agglomeration Externalities and Industry Dynamics in Empirical Studies Literature

5.2.1 Dynamic Externalities Based on US and EU Regions' Experience

The previous chapter emphasizes the role of knowledge spillovers as crucial sources of innovative activity in the high-tech industries. It has also underlined that knowledge spillovers are most likely to occur in urban areas. Due to their concentration of economic activities, urban areas create positive dynamic externalities of agglomeration,[1] which, in the words of Glaeser et al. (1992, p. 1127), allow "(...) to learn from others and thus improve one's own productivity (...)." The latter explains the advantages of clustering from the perspective of the clustered firms and institutions. The presence of dynamic externalities explains both the formation of urban areas and a local economic development.

A detailed description of the three major theories that explain the origins of dynamic externalities is contained in the Marshall-Arrow-Romer's (MAR), Jacobs's and Porter's theories and was presented in Chap. 1. This section explores key empirical evidence on the dynamic externalities found for the US and EU regions.

Glaeser et al. (1992) use data on US cities to estimate the effects of local industrial structure on growth. The unit of observation is the couple sector and the city, of which he observes the level of employment over 2 years (1956 and 1987). He takes into account the largest cities only and the six largest-digit industries in each city. Furthermore, he considers the employment rate as a function of specialization, diversity, and competition indicators, as well as other control variables. His results show positive effects of diversity and competition, but no significant effect has been found for specialization, which is in line with Jacobs and Porter's theories but opposite to the MAR externalities.

Feldman and Audretsch's (1999) and Kelly and Hageman's (1999) studies provide evidence of the increasing role of diversification of externalities for

[1] Clustering of economic activities may also lead to negative externalities such as congestion and pollution that can offset the positive economic outcome of the clusters.

regional innovative activity in the US. A number of studies provide evidence for both types of externalities based on the experience of a smaller sample of European NUTS2 and NUTS1 regions (Caniels 1997; Bottazi and Peri 2000; Moreno et al. 2003; Greunz 2005; Fischer et al. 2005; Rodriguez-Pose and Crescenzi 2006; Maggioni et al. 2007; Crescenzi et al. 2007).

Caniels (1997) found clear evidence that spatial concentration in sectors effects the spatial concentrations of innovative as well as economic activities of the NUTS1 and NUTS2 regions. This indicates that industry-specific as well as locally originated conditions determine the spatial pattern of innovation.

Bottazi and Peri (2000) attempted to identify and estimate a "production function" of innovation for European 86 NUTS1 regions using patent and R&D data for the 1977–1995 period. They found that knowledge externalities exist within geographically localized actors; however, they didn't give any evidence of the role of technological space for knowledge externalities, thus no MAR or Jacobs externalities were identified.

In analyzing the European landscape on the NUTS1 and NUTS2 levels on the sample of 175 regions,[2] Moreno et al. (2003, 2005) provide evidence for the MAR externalities. Moreover, the authors aimed to capture the differences in technological and economic structures across the EU regions in terms of manufacturing employment in each region, biotechnology, organic chemistry, and pharmaceuticals and their patenting activity, according to the IPC classification. In general, the results point to the existence of MAR externalities. However, the results for the biotechnology and pharmaceutical industries show a negative correlation between the share of patents and innovations, giving more credibility to Jacobs's externalities.

In her study, Greunz (2003, 2005) tested the European landscape based on 158 NUTS2 for the MAR, Jacobs, and Porter externalities in four technological groups: high technology, medium-high technology, medium-low technology, and low technology. Her findings suggest that Jacobs's externalities have a considerably higher influence on knowledge creation than MAR externalities have, although both kinds of externalities do have an impact. These results suggest major innovations in the high-tech industry are not inspired by intra-industry technological solutions, but come from an entirely different industry. Similar results were achieved by Henderson and Cockburn (1996) for the US case; they concluded that high-tech industries are more likely to locate in places with a history of industrial diversity.

The study by Crescenzi and Rodriguez-Pose (2006) suggests a mix of MAR and Jacobs externalities in the EU regions (NUTS1 and NUTS2), when a broad group of manufacturing sectors was considered. The authors state that although agglomeration is a better driver of innovation, regional specialization is negatively associated with the innovation process in the European regions. Furthermore, they conclude

[2] The empirical analysis is conducted for the NUTS 2 regions of 17 Western European countries (EU-15 plus Norway and Switzerland). However, NUTS 1 regions have also been considered for the UK, and NUTS 0 for Denmark and Switzerland, due to data constraints.

that agglomeration externalities in the EU regions are severely limited by the EU's low level of factor mobility. Similar results were achieved by Van Oort (2002) (on The Netherlands) and Ouwersloot and Rietveld (2000) (on the US). They emphasize the importance of diversification externalities for innovation in manufacturing industries, as do Kelly and Hageman (1999). Using R&D labor costs data for The Netherlands (Van Oort 2002) and US patent data (Kelly and Hageman 1999), they highlight the importance of diversification externalities for innovation in manufacturing industries. Kelly and Hageman (1999) observe that "the location of Research and Development (R&D) centers is determined more by the location of other sectors' innovation rather than by the location of its own production."

Contrary to Crescenzi and Rodriguez-Pose (2006), the study results by Maggioni et al (2007) confirm the significant impact of the relative specialization in high-technology sectors on the innovation activity of the EU NUTS 2 and NUTS1 1 regions. However, their results are too general and vague to assume that an increased concentration of firms of the same industry within a region facilitates any intra-industry knowledge spillovers. On the other hand, Paci and Usai (2000) show that in the European regions there is a positive association between the spatial distribution of technological activity and production specialization. This result has been confirmed by a more robust econometric analysis conducted by Greunz (2005), and based on a larger sample of countries. The author found a positive and significant influence of the specialization measure on the patenting activity of the high-tech industry and gave support to the possible existence of MAR and Porter externalities in the EU-15 regions.

Kelly and Hageman (1999) applied state patent counts data to their model to track knowledge spillovers. They found that patenting was geographically clustered. A disadvantage of their study may stem from the fact that they consider the state as a geographical scale for research. However, the state is not an appropriate geographical unit for measuring knowledge spillovers, because the state's administrative boundaries may not be the same as the boundaries of the economic interactions they hold (Breschi and Lissoni 2001). Moreover, US metropolitan areas sometimes extend across state boundaries.

The results of Baptista and Swann (1998), for the sub-national level, support the fact that firms in well-developed industrial clusters innovate more actively. They apply data they collected from 248 firms in the UK between 1975 and 1982. Their findings suggest a positive correlation between a firm's innovativeness and the region's employment in the same industrial sector. Baptista (2000) found that the presence of early adopters of new technology is positively correlated with the overall efficiency of knowledge flow in that region. While investigating the performance of companies in the ICT sector, he found that the 'number of previous adopters of technology in the region' is positively and significantly correlated with present determinants of adoption of microprocessors and computer numerical control in the region. His study confirms evidence of local knowledge spillovers through correlations between innovativeness of a region or a firm and that of its surrounding area. While acknowledging the value of these findings, it should also be pointed out that this evidence is indirect to say the least. Due to the aggregate nature of his data, he could not study individual points of knowledge spillovers.

5.2.2 *Porter, MAR, and Jacobs Externalities and Industry Life Cycle*

A limited number of studies provide evidence of the relationship between industries' life cycles and agglomeration externalities. They include studies by Neffke et al. (2008); Duranton and Puga (2001); Lundquist and Olander (1999); and Nelson and Winter (1982). Broadly, they conclude that a part of the variation in externality is determined by differences among industries, whereas some externalities may be general throughout the whole industrial development process. Considering a general industry life cycle model, most of the authors argue that development of innovative products takes place in big and diversified productive structures, whereas the production process and standardisation take place in smaller, specialized locations (e.g. Lundquist and Olander 1999; Duranton and Puga 2001). For example, Neffke et al. (2008) investigated the Swedish regions between 1990 and 2004 and found that benefits industries derive from their local environment are strongly associated with their stage in the industry life cycle. According to the authors, firms and agents of industries engage in competition based on quality differences in product varieties. A diverse industrial structure increases the flexibility of young industries by enlarging the variety of input alternatives (Neffke et al. 2008).

One of the most relevant findings of Henderson and Cockburn (1996) is that high-tech industries are characterized by positive specialization and diversity effects, which is in line with the presence of both MAR and Jacobs's externalities. The author concentrated on the effects of diversity and specialization on the evolution of industries (but excluded competition from the analysis). According to the author, these findings are consistent with the concept of industry life cycle. New industries emerge primarily in large and diversified urban areas, but when they mature, they decentralise their production to smaller specialized cities. Perhaps the most striking finding is that during the mature and declining stages of the life cycle of an industry, increases in the geographic concentration of production tend to lead to greater and not less dispersion of innovative activity. This suggests new ideas need new space, at least during the mature and declining stages of the industry life cycle. In any case, the positive agglomeration effects during the early stages of the industry dynamics are apparently less important during the latter life cycle stages (Feldman and Audresch 1996).

Considering this limited number of empirical studies, there seems to be little or no common consensus in the literature concerning the relationship between the above-mentioned externalities and industry dynamics. This calls for more empirical evidence to establish whether MAR specialization or Jacobs's diversification externalities favor regional innovativeness, and in which high-tech industries, as well as what the role of technological maturity of these industries is in explaining the presence of externalities in regions.

5.3 Patent Activity and High-Tech Industries Maturity

One of the generally adopted approaches towards studying the technological maturity of industries is tracking the intensity of patent applications. Several researchers, including Haupt et al. (2007); Watts and Porter (1997); and Andersen (1999) have tried this approach. Patents embody information about technological development itself since they contain the technological know-how. Registered patents inform us about research intensity and innovativeness, and therefore the technological state of the art in the relevant field. The number of patent applications can vary at different stages of the technology life cycle. At the *emerging* and *growth* stages, the indicator of the number of patent applications is typically higher than that of all other indicators. When basic technological and market uncertainties have vanished, innovations become less radical, R&D risk decreases and the number of patent applications may increase. In the following phase of *maturity,* the number of patent applications (typically incremental innovations) remains constant. After that, when the potential for new product innovations decreases on the basis of technology, and consequently the number of annual patent applications decreases constantly, the technology's *decline* stage begins.

However, patents can only affect industry dynamics if the market adopts them. In other words, when the new technology is widely diffused and used. Within this context, adoption refers to the stage at which individuals or organizations select the technology for use. Diffusion refers to the stage at which the technology spreads to general use and application.

Based on the studies of Wunderlich and Khalil (2002) on technology diffusion and maturity analysis as applied to different industrial sectors, and biotechnology based on Utterback and Abernathy (1975) and Fisher and Pry (1971), the following section discusses the maturity of technology and its relationship to technology diffusion. The measure of technology diffusion may be expressed as a percentage and can be characterized over time by applying the Fisher-Pry equation. Two characteristics of the Fisher-Pry equation are that by measuring technology maturity through technology diffusion – L/(1 + e**(−b(t − a))) it firstly reveals the midpoint in time (identified as "a") at which the market achieves 50 % adoption, and secondly, provides a relative quantification of the rate (known as "b") at which the technology was adopted. Based on data of the US National Academy of Engineering (NAE),[3] data on the timeline summarizes the 20 most notable

[3] Greatest Engineering Achievements of the twentieth Century, www.greatachievements.org/greatachievements

technological engineering achievements, and Wunderlich and Khalil (2002) have grouped them into industrial sectors accordingly. The technology performance parameter extracted from these timelines is the technology diffusion parameter. The timelines of these technologies enable at least a rough estimate to be made for the start of the rapid growth of each technology. The scores ranged from a 0 % to 100 %, based on the Fisher-Pry equation for technology diffusion. Since the GEA data is for the twentieth century, an initial score was established for the year 1900 (Wunderlich and Khalil 2002). A score of zero would indicate that the technology had not yet been conceived. For example, the start of technology for laser and fiber optics moved from a base level of zero to a score of 1.0 when Einstein established the theory of stimulated emission in 1917. More significantly, a score of 15 was assigned as the milestone indicating that rapid market growth was about to occur. It is assumed that at this point anyone who wanted to obtain the technology could get it. A classic example of a 15 % score would be the rise in automobile technology in 1908, when mass production of the Model T Ford began. Then, the area between 1 and 15 was feathered, based on the early developmental milestones listed, such as the origination of mass production techniques by Ransom E. Olds in 1901. The significance of the 15-point milestone for technology diffusion is that it indicates that rapid market growth has been initiated and will carry on until it reaches another significant point of 85 %, which indicates the end of rapid market growth. At this point of time, the technology is judged to have achieved widespread diffusion in a way that nearly everyone who had wanted to obtain it had done so (according to Wunderlich and Khalil 2002). A score of 100 would indicate that a specific technology had reached complete diffusion or maturity.

On the basis of this more selective rather than exhaustive survey, the following section discusses technological maturity of each high-tech industry as a proxy for industry dynamics. The results of Wunderlich and Khalil (2002) and Utterback and Abernathy (1975) (for biotechnology industry findings) for each high-tech industry are illustrated in Fig. 5.1.

As shown in Fig. 5.1, the *Aerospace industry* (both the aeroplane and spacecraft industries) is currently in the growth stage of its life cycle. In general, the industry has extremely long product life cycles and development periods. The aerospace industry is characterized by its renowned heavy upfront investments and exceptionally long program lives. It normally tends to be very regulated and also requires huge stocks of existing knowledge (even though a slow deregulation of the industry in recent decades has lowered entry barriers into its market). As a result, the process of technological substitution and adoption of new products and processes is very long indeed. In technical terms, the technology diffusion for the aircraft manufacturing industry has achieved a high rate of maturity, because now anyone who can afford the products in that market is able to obtain them without difficulty. However, because of a big lack of expertise of suppliers and users of specific technological processes, the diffusion process may be an "incomplete success or at times a failure on the part of the user to completely absorb the technologies provided by the suppliers" (Rasheed and Manarvi 2008). Some smaller firms may be disadvantaged due to inadequate human resources, and weak organizational and R&D capacities. As a result, the dynamics of the aeroplane industry may be slow;

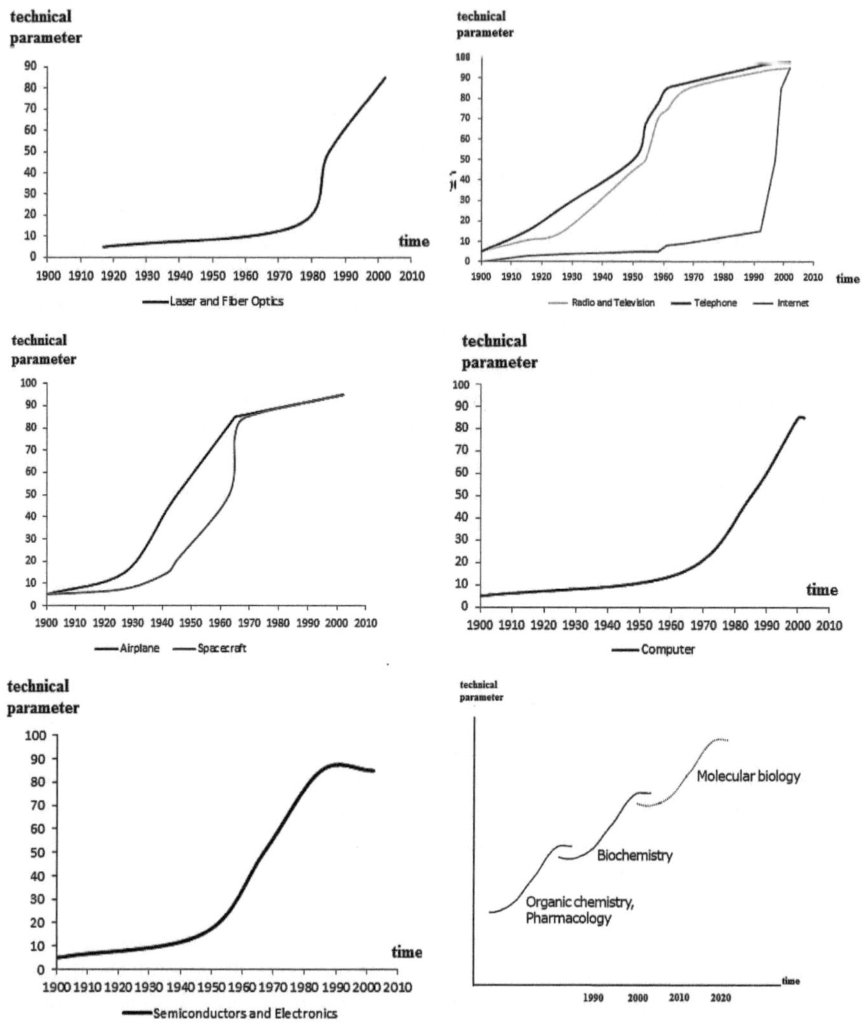

Fig. 5.1 Industry life cycle maturity and technology diffusion in high-tech industries. Note: Technology Life Cycle Maturity Level Score: Solid concept/idea conceived 1; Rapid growth enabled 15 Peak growth; 50 % diffusion 50; Rapid growth slows, market-saturating 85; Market saturated, technology mature 100 (Source: Own elaboration based on Wunderlich and Khalil 2002; biotechnology based on Utterback and Abernathy (1975) and Fisher and Pry (1971))

however, it is far from its maturity stage. Because of the high investments into R&D and high cost-related activities, it seems the industry is fairly concentrated.

Furthermore, industry dynamics converge with the expansion of other industries. IT investments are becoming increasingly important. In the airline industry computers and IT in general have both become a necessity for many parts of operations and marketing (Subramanian et al. 1994; Anbil et al. 1991).

Additionally, the industry has been facing a gradual deregulation of competition.[4] The recently noted consolidation in the form of mergers between companies or takeovers of failing carriers may be a part of the birth-and-death process expected to be in existence in a competitive marketplace such as the airline industry (Swann 2005).

In summary, the aerospace industry still offers the most attractive competitive opportunities with its products and services (Porter 2008) and proves that it is still in the growth phase of its life cycle. Thus, the role of diversity and heterogeneity plays a key role and may also have a significant impact on the industry's competitive advantages and further dynamics.

The development of the *Computer industry* illustrates the traditional industry life cycle by first experiencing product innovation, then processes and services innovations. In the old days, when mainframe computers dominated the industry, researchers and engineers paid more attention to product or system design issues than to the process of constructing a particular piece of software or an operating system. Based on the Fisher-Pry score, Fig. 5.1 shows that in the mid-1980s computer technology was at its highest growth rate (with the PC industry as its driving force). It retained its high growth rate for over two decades, mainly due to upgrades in hardware, services, and add-on products and features. By the end of 2000, the market became very competitive and as Fig. 5.1 suggests, the industry entered its early period of the maturity stage. This trend could be supported by observing the present market, with many manufacturers producing a range of models, from desktop computers to notebooks, and continuing the product differentiation process which began in the growth stage.

Despite these slowing dynamics of the computer industry in highly advanced market economies such as the US and Europe, the diffusion of computing technology and its adoption around EU countries was proven to be significantly various, depending on the region concerned. The results obtained in 2007 reveal a significant difference in household computer ownership between the European Union's old and new Member States: EU-15 (60 %) compared to EU-12 (45 %) (Eurobarometer 2007). The reasons for the slower computer adoption in the EU-12 could be associated with income levels, qualifications of human capital, investment rates, property rights protection, and a big share of agriculture in their GDP.

In sum, the computer industry shows slowing dynamics or greater stability over the past decade, but because of the diffusion of computing technology and its adoption in advanced countries, the effects of local industrial structure on industry growth may differ from country to country.

[4] In a 1981 study, David Mowery and Nathan Rosenberg have argued that rapid diffusion of technological innovations in the U. S. commercial aircraft industry to U. S. airlines during the mid-twentieth century was due partly to actions of regulatory agencies, http://elsa.berkeley.edu/~bhhall/papers/HallKhan03%20diffusion.pdf.

The *Communication industry* is one of the prime support services needed for the rapid growth and modernisation of various sectors of the global economy. It has become especially important in recent years due to the enormous growth of information technology (IT) and its potential. In fact, the biggest drivers of the communication industry are the availability of broadband infrastructure and a speedy rollout of new services, which may, however, be slowed down by regulations over privacy and security issues that have been introduced. Meanwhile, ownership of a telecom license can represent a huge barrier to entry into the communication market. In addition, solid operating skills and management experience is fairly scarce, making entry into the industry even more difficult.

Broadly, the communication industry sector is composed of three main sub industries: radio and television, telephone, and Internet industries. Based on the finding of Fisher-Pry, technology diffusion in the telephone, radio, and television sectors is moving closer to maturity, scoring 98, 95, and 95 respectively. The results of this Fisher-Pry-based analysis of the telecommunication industry are not in coherence with general observations that have been carried out regarding the EU telecom market and Internet technology diffusion, in particular. In terms of EU experience, the development of telecommunication technology has been rapid in recent years, with IT and related services playing important roles. The telecom sector has experienced a major process of transformation in terms of growth, technological content, and market structure in the past decade through friendly policy reforms. Roughly, the EU completed the liberalization of the European telecommunications market in 1998.[5] As reported by Eurobarometer (2007) study the penetration of telephone access in EU-27[6] was 95 % by mobile phone, fixed phone, or both. The penetration rates in the EU-15 Member States appear to be significantly higher than in the case of the EU-12 (97 % compared to 91 %). The mobile phone penetration rate among households in the EU-27 is 83 % (84 % in the EU-15; 78 % in the EU-12). Also, in terms of access to television, the penetration rate in the EU-27 is 96 % (Eurobarometer 2007). Internet technology diffusion in the EU-27 is still in its rapid growth stage. The proportion of households in the EU with access to the Internet reached 73 % in 2011 compared to 49 % in 2007.[7] Internet access, however, remains considerably higher in the EU-15 (73 % (52 % in 2007)) than in the EU-12 (below 60 % (33 % in 2007)) (Eurostat 2011; Eurobarometer 2007).

Even if the communication industry could have been considered as an industry reaching its phase of maturity, Internet technology and the diversification of

[5] The agreement on the timetable for full liberalization included transitional periods for certain Member States. As a result of a case-by-case assessment by the Commission, the following periods have been confirmed: Luxembourg fully liberalized its market from July 1998; Spain from December 1998, Ireland and Portugal from January 2000, and Greece from January 2001. For Member States joining the EU in 2004 and 2007, the liberalization process was completed at a later date.

[6] Including the two latest Member States, Bulgaria and Romania.

[7] More widespread in metropolitan areas than in rural regions.

products and services of its application will continue creating new market opportunities for the telecommunication industry.[8] New Internet applications emerge and prosper in diverse fields such as electronics, medicine, biotechnology,[9] aerospace, etc., which suggests that a diversified local industrial structure could generate the biggest technological externalities for this industry.

The *Semiconductor industry*[10] has been reporting constant growth since it came into being in the late 80s and through the 90s. Since its beginning, the semiconductor industry has continually introduced innovative products. According to Gruber, the adoption of new technologies is extremely rapid (1994). Declines in the price/performance ratio of semiconductor components have propelled their adoption in an ever-expanding array of applications (in electronics industries; telecommunications, automobiles, military systems, consumer electronics, personal communications, and home appliances) (Macher et al. 1999). However, Gruber (1994) points out that even though semiconductor manufacturers receive a great deal of publicity, they make their money mainly from semiconductors in their growth and maturity stages. The past decade has observed the maturing process of the semiconductor and electronics industries (based on a Fisher-Pry score for technology diffusion). The decline of the semiconductor industry is explained, firstly, by the fact that there are no immediate substitutes for semiconductor chips. Therefore, current substitution is extremely low. Secondly, by the competitive nature of the marketplace and the need to acquire production experience and move down the learning curve (Gruber 1994). In fact, innovations in the semiconductor industry depend highly on the stock of existing knowledge of productive units and the process of 'learning-by-doing spillovers.' Learning-by-doing spillovers may also generate a string of incremental innovations and may perhaps allow firms to enter the desired martkets (Malerba and Orsenigo 1994). Furthermore, as many authors emphasize, semiconductor firms focus on the improvement of quality and thus vertical product differentiation (Malerba 1985, Georghiou et al. 1986). The latter requires specialized human capital (scientists, engineers, developers, etc.), which is not easy to duplicate. All this suggests the biggest technological externalites will result from the specialized local industrial structure, possibly clustered around universities and corporate R&D centers.

---

[8] The Internet has already had a strong effect on the financial services, telecoms media, and government (eGovernment services, etc.), and has an even greater impact on the coordination and integration of all manufacturing operations across different geographies e.g. through Internet-enabled supply chains, eCommerce and eBusiness applications (Gaffard and Krafft 2000).

[9] Bioinformatics is a branch of biotechnology that uses information technologies to work with biological data like DNA. http://www.careercornerstone.org/industries/sciresearch.htm

[10] Semiconductors are classified into major product groups, mainly according to their function. The largest product group is memory chips, which account for about one third of the total semiconductor market. The second-largest semiconductor product group are micro components, which consist mainly of microprocessors and micro controllers. The remaining part of the semiconductor market is made up of logic devices (including application-specific integrated circuits or ASICs), analogue devices, and other parts (Gruber 1994).

Based on the theory of the innovation life cycle, the process of technological change in the *Biotechnology industry* represents technological evolutions in the biopharmaceutical industry, as a whole. Biotechnology is a relatively young branch of bioscience, developed by the biopharmaceutical industry in the late 2000s. The innovation process shows that there is not just one S-curve but a succession of S-curves from organic chemistry/pharmacology to biochemistry and molecular biology (Fig. 5.1). It can be seen that the waves of molecular biology overlap the waves of biochemistry and are about to leap upwards, according to Utterback and Abernathy (1975). Currently, scientists and researchers are attempting to exploit basic molecular research to identify new drugs, the production of which will be based on recent advances in genomics technology. Scientific breakthroughs such as genetic engineering, the ability to create monoclonal antibodies, and the mapping of the human genome have opened up new areas of research, and the pace of discovery in basic biomedical science has accelerated dramatically over the past few decades.

The emergence of biotechnology is changing the pharmaceutical industry in terms of requiring a convergence of science and technologies and a multi-disciplinary approach to produce new technological discoveries (biological sciences, chemical engineering, bioprocess engineering, information technology, biorobotics). Increasing competition drives the specialization of firms in specific products; however, so far this has been somewhat limited due to the few experts in the specific biotechnology fields, e.g. cancer diseases.

Biotechnology is firmly rooted in the growth stage, with heavy reliance on science and R&D investments. Patenting has increased sharply over the past few decades,[11] with biotechnology patenting applications far outpacing the general rise in patenting applications. The biggest number of patents in biotechnology grew from the late 90s up to early 2000. For example, in 1977, there were only 12 biotechnology patents filed globally under the PCT. By 2009 this number had increased up to 9,339 patents (this is substantially more than a 77 % increase). Almost 70 % of these patents were filed by an inventor resident in either the EU-27 or US (Patent statistics, OECD 2012). However, while the biotechnology sector shows a strong growth stage, the degree of diffusion and adoption of biotechnology products and processes has been slowed down for several reasons. Although product innovations enabled by biotechnology have increased the quality and variety of goods and opened up new markets, integrating product innovations into modelling frameworks is difficult according to Pianta (2005). Furthermore, the substitution of traditional techniques for producing products with the use of biotechnology is related to the costs of transformation of existing production processes, e.g. substituting diesel extracted from petroleum with biodiesel made from

[11] Still, this is not to say that biotechnology patenting outside the OECD did not experience significant growth. In fact, in the BRIC economies, as well as a number of other Asian and Latin American 'tigers,' biotechnology patenting increased substantially over the same time period.

feedstock or canola oil (McNiven 2007). Moreover, in the present case of biopharmaceuticals, the demand side is largely influenced by regulations. It is strongly regulated and therefore excludes many inventions due to morality (based on Art. 53(b) of EPC). The latter may have an impact on further intensity in biotechnology. These influence the financing of new products and the degree to which markets may grow. As a result, industrial biotechnology is still in its relatively early stage of growth, and many potential products are not yet on the market.

The level of activity in the biotechnology industry among the EU countries depends largely on the research field. Europe's competitive edge lies mainly in healthcare applications and in industrial biotechnology, including the chemical industry. Some Member States have developed advanced biotech sectors whereas others have stayed behind (Denmark, Germany, UK). New Member States of the EU are generally the early movers in the biotechnology sector. Thus, the identification of the stage of life cycle of the biotechnology industry must be treated with necessary caution.

In summary, the analysis of the growth stage of biotechnology shows that the industry is still in the initial stage of growth in its life cycle. This requires huge amounts of R&D funding, whereas R&D projects often involve high risk of failure. Biotech firms that are active in the biopharmaceutical sector and do not have alliances with large pharmaceutical firms, tend to rely more heavily on domestic sources in their innovative activities, including universities and public research organizations.[12] Thus, in order to stimulate positive technological externalities in the biotech industry, the local productive structure must be determined by the presence of diversified local technological capability centers (clustered near universities).

The laser technology industry is the sector showing characteristics of an industry still in the early phases of its evolution. A huge variety of lasers has been developed since they were first invented 50 years ago (Fig. 5.1). Laser systems prove to be a core technology, enabling many new applications, particularly in medicine. They can be available in many forms, each of them being of a different nature and having a different purpose. Laser applications are equally varied and are employed, for example, in manufacturing (e.g. welding and marking), surgery, defence (e.g. ranging and weaponry), scientific research, and development. The rich applicability of laser technologies in a wide range of disciplines suggests that their industrial cluster is very heterogeneous and diverse.

The annual number of patent applications has decreased (in 1998–2001 there were on average 50 patents per million inhabitants in the EU, whereas in 2002–2008 this number dropped to 31 patents, on average). This may indicate that the innovation system in the field of laser beam sources has reached a certain stage of maturity or that its growth is slowing down.

---

[12] www.ndu.edu/icaf/programs/academic/industry/reports/2011.

Similar results are presented in Fig. 5.1, which illustrates the Fischer and Pry score for the laser and photonics industry and shows that the rapid growth of both industries slows down as the market for their applications becomes saturated (85). The dynamics of innovation and diffusion processes could be infringed by several factors. The first is related to the complexity of the technology. Each application for laser technology can be thought of as a technological opportunity requiring a range of different patentable inventions that are combined in a functioning product. In another words, a product using laser technology will usually also embody some patents relating to different technological areas outside of optics (Graevenitz et al. 2011). Finally, laser technology industry experiences long technology cycles, so the process of substitution and adoption may take longer, too.

5.4 Identifying MAR, Porter, and Jacobs Externalities for the EU Regions

5.4.1 Framework and Model

There are a number of important structural differences between EU-10 and EU-15 Member States' regional economies concerning production systems. One of these differences was already explained in Chap. 3 and originates from the different historical and institutional backgrounds of their 'innovation systems,' whereas the second one stems from the rather vibrant business structure of the EU-10 countries. The decentralization and transformation processes of the EU-10 countries resulted in a sharp rise of the number and kinds of organizations present in the economy. Two decades ago these economies were dominated by a relatively small number of large organizations, such as state-owned enterprises and public research institutes that were characterized by a high degree of internal division and specialization. Innovation in a certain field was often concentrated in one or a few organizations only. This is especially true for high-tech industries, where intensity of competition almost didn't exist. As a result, high-tech and medium high-tech sectors were less competitive in the EU-10 regions. Since the period of analysis covers the period of the transitions (1998–2007), it is important to include the technological paths of the EU-10 regions in the context of the current local productive systems of the EU regions (Fritsch and Werker 1999).

The following conceptual model uses the MAR, Porter, and Jacobs externalities framework to test whether the externalities arising from industrial diversity as well as externalities resulting from industrial specialization positively influence knowledge creation capacity in the EU regions. Since regional technological externalities may differ depending on the industry that is being examined, the model is tested separately for each of the six high-tech industries: Computer and electronics, Microorganizm and genetic engineering, Aviation, Communications technology, Semiconductors, and Lasers. Furthermore, the model accounts for sector-specific characteristics such as technological opportunity, overall dispersion of relevant employment in high-tech industries among European regions, as well as a region's specific characteristics, such as its size and endowment of knowledge sectors etc., which is assumed to positively influence patenting activity of the concerned region.

Since the dependent variable of the model has a discrete nature with an important proportion of zeros, the use of conventional linear regression models may be inappropriate. Generally, in order to deal with the discrete and non-negative nature of the patent-dependent variable, the simple Poisson regression model is considered as it was in the study of Feldman and Audretsch (1999). However, a significant shortcoming of the Poisson model is its implicit assumption of equality between the first two conditional moments.

Taking into consideration the evidence of over-dispersion and the rejection of the Poisson restriction, the model is estimated by allowing for a mean-variance inequality. The adopted approach consists of applying the Count Models for panel data: the Poisson quasi-maximum likelihood estimator (QMLE).[13] It is worth noting that the interpretation of the estimation results of non-linear models such as the Poisson, Tobit, or negative binominal model, is somewhat different from those obtained by linear regression models. However, when the explanatory variables enter logarithmically, then the coefficients can be interpreted as an elasticity giving the percentage change of the expected value of the dependent variable for a 1 % change of an explanatory variable. For this reason the independent variables of the equation have been expressed in logarithmic form.

$$P_{ij} = a_1 + a_2 HTKS_{ij\,t-3} + a_3 S_{ij} + a_4 COMP_{ij} + a_5 EKC_{it-3} + a_6 POP_i + a_7 GD^{P}{}_{it-1} + a_8 GD^{K}{}_{it-1} + a_9 RDi + \varepsilon_{ij}$$

Where,

i region
j industry
Pij stands for the average number of patent applications over the 1998–2007 periods in region *i* and sector *j* to the EPO and proxies innovation output.

5.4.1.1 Specialization Measures

As the previous section emphasized, the importance of agglomeration economies, the presence of cluster, and intra-industry transmission of knowledge seem relevant in fostering technological externalities in some of the high-tech industries. It is, however, crucial to verify in which industries agglomeration economies lead to

[13] This robustness of the Poisson QMLE is a special case of the more explicit analysis of quasi-maximum likelihood estimation in the linear exponential family by Gourieroux, Monfort, and Trognon (1984).

higher regional innovative performance (given their specific sectorial characteristics $HTKS_{ij}$ *and* S_{ij}).

$HTKS_{ij}$ regional production specialization in *high-tech* knowledge-intensive services 1998–2007

S_{ij} regional production specialization in *high-tech* industry (j) 1998–2007

Measures of regional specialization and spatial concentration are calculated on the basis of regional employment data on manufacturing branches. For each region, the specialization index is calculated on the basis of the regional share of employment in a given sector k.

$$S_{ij}^{P} = \frac{E_{ij}}{\sum_{j=1}^{n} E_{ij}} \Bigg/ \frac{\sum_{i=1}^{m} E_{ij}}{\sum_{j=1}^{n}\sum_{i=1}^{m} E_{ij}}$$

where

i indexes the region ($i = 1\ldots m$), m = ...;
j indexes the high-tech sector employment ($j = 1\ldots n$), n = ...7;
E stands for average employment over the period 1998–2007 (Source of data: Eurostat – REGIO).

If the degree of industrial specialization of a given region equals the European average, the value of the indicator is one, while a higher/lower value points to a higher/lower degree of specialization. The lower bound of the index is 0.

5.4.1.2 Diversity Measures

For many other industries, according to the presented arguments in the previous section, local diversity may be the major engine for fruitful knowledge spillovers and technological externalities. In order to contribute to the existing discussion on the role of industrial diversity in generating technological externalities for specific types of high-tech industries, the reciprocal of the Gini coefficient is applied (Greunz 2004).

GD^{K} *i* production diversity indicator of a region (in high-tech sector) in 1998–2007

$$GD_{i}^{k} = \frac{2}{(n-1)\sum_{j=1}^{n} E_{ij}} \sum_{j=1}^{n=1} CE_{ij}$$

where the definitions of *i*, *j*, and *E* are the same as in the Sij index, and where *CEij* is the cumulative sum of all employees in the industrial sector *j* when sector employment is listed in increasing order.

The reciprocal of the Gini coefficient varies inside the interval of (0, 1) and increases together with production diversity.

Complementarily to production diversity, the author tests for the impact of innovation diversity. The applied measure is the same as the reciprocal Gini coefficient applied to the local diversity index (GD^K *i*), except that instead of employment, patent data is used. Innovation diversity is therefore defined as follows (Greunz 2004):

$GD^P{}_i$ innovation diversity indicator of a region (in high-tech sectors) in 1998–2007

$$GD_i^P = \frac{2}{(n-1)\sum_{j=1}^{n} P_{ij}} \sum_{j=1}^{n=1} CP_{ij}$$

where the definitions of *i* and *j* are the same as previously. *P* refers to the average number of patent applications over the period (1998–2007) and to the EPO attributed to the living place of the inventor. *CPij* is the cumulative sum of patents within the industrial sector denoted as *j* when sector patents are listed in ascending order. A positive and significant sign on the coefficient of innovation diversification is interpreted as evidence of the presence of Jacobs's externalities.

The impact of local productive externalities on the innovative performance of given regions depends largely on other characteristics, such as technological opportunity/intensity, competitive conditions, region-specific measures (size of a regional economy and of education and R&D-based activity), industry life cycle, and other industry-specific variables, e.g. strategies and activities of innovating firms. The following model cannot capture all of them due to statistical and methodical limitations, but it will try to consider the most relevant ones for the purpose of this chapter.

5.4.1.3 Competition Measures

The competition intensity in the high-tech markets may be discouraging for any potential entrants. Changes in the number of competitors (both globally or nationally) and their strategies affect the region's technological externalities and their innovative performance as a whole. Competition is captured by measuring the number of establishments per employee (COMP) in the site-industry, relative to establishments per employee in this industry on the overall industrial area of the region. where F denotes the number of firms.[14]

$COMP_{ij\ t\text{-}3}$ competition index in a specific high-tech industry in region 1998–2007 of region i and sector j

$$\mathrm{COMP_{s,i}} = \frac{\mathrm{F_{s,i}/E_{s,i}}}{\sum_{s=1}^{m} \mathrm{F_{s,i}} / \sum_{s=1}^{m} \mathrm{E_{s,i}}},$$

A value greater than 1 means that the specified industry contains more firms relative to its size on an industrial site vis-à-vis the total amount of industrial area in the specific region. Glaeser et al. (1992) claim that a value greater than 1 can be interpreted in a way that the industry on a site is locally more competitive than it would be on a site elsewhere.

5.4.1.4 Region-Specific Measures

The share of employment in the education and knowledge-creation sectors, the total R&D expenditure per capita, and an average population size in a given region (which may, by the way, shape the innovative performance of selected regions) all act as a proxy for the underlying economic structure.

EKCi employment in education and knowledge creation in region *i* in 1998–2007

RDi identifies the total R&D expenditure per capita in region *i* as PPP in 1998–2007 (at constant 2,000 prices)

POPi is the average population size over the period 1998–2007 introduced as a size control variable;

eij denotes a random error.

Furthermore, since MAR as well as Jacobs externalities are considered to be the most compelling in the context of large agglomerative areas, a dummy variable of the 50 most densely populated regions is introduced in order to capture the direct impact of having a higher population density on the number of patents. According to Ciccone (2002) and his studies, population density could be a proxy for agglomeration economies. High population density increases the chance of finding positive effects of specialization and diversity on innovative output. It is argued that the spatial concentration of individuals, capacities, information, and knowledge within a limited geographic area provides an environment in which ideas flow quickly from person to person and are exchanged rapidly between institutions. In other words, dynamic technological externalities arise from communication between economic agents and the number of R&D institutions, which is disproportionally bigger in larger cities.

---

[14] The application of this measure is in line with Glaeser et al. (1992), who consider the number of firms per worker as a proxy for competition. Glaeser et al. (1992) reason that a value greater than 1 can be interpreted that the industry on a site is locally more competitive than it would be on a site elsewhere.

5.4.2 Estimates Results

It was the author's aim to test the model on as many EU regions' productive structures as possible; however, due to limitations in statistical data (mainly for the EU-10 regions) as well as poor patent performance in all six types of high-tech disciplines, the numbers of observations have been cut for each of the high-tech industry groups. The samples of regions include 25 regions for the aerospace industry, 142 for the communication industry, 142 from computer science, 120 from semiconductors, 149 from biotechnology, and 156 for laser industries. The observations were based on the availability of patents data. Countries with very few high-tech patent applications did not record patenting activity in all the groups. The estimate results for each industry presented in Table 5.1 were achieved based on the Eurostat and national statistical offices data for the 1998–2008 period.

The outcome for the *Aerospace industry* suggests that production diversity variable GD^k turns out to be a particularly important input for new knowledge creation in the a*erospace* industry. According to estimates results, a 1 % increase of production diversity GD^k of an average EU region increases the number of patents by about 7,706 % (with 99 % significance). The latter findings roughly confirm that the EU aerospace industry is still at a growing stage of its technological maturity and industry dynamics convergence with the expansion of other industries, especially IT.

Furthermore, the estimates presented in the table indicate that a 1 % increase of employment in the education and knowledge creation sectors of an average *EKC* EU region induces a 3.94 % increase in the number of patents in a*erospace* (with 95 % significance). Thirdly, when investigating the relationship between patenting activity and the size of regions, measured by number of population *POP*, one would expect a number of patents to grow disproportionately in larger urban centers, thus showing increasing returns in inventing activity with respect to population size. One of the striking results of the model is that the number of population may have a slightly negative effect on patenting activity in the aviation industry. Along with the increase in population size by 1 %, the number of patents in aviation drops slightly by −1.51e-07 (with 95 % significance).

Altogether, this may also suggest that what matters for innovation in the aerospace industry is not simply the presence of an economy with a large number of people, but more importantly how its technological capabilities are integrated by the environment they create and live in. This argument may be taken to point to specific characteristics of a region, beyond its population size. Public institutions for procurement, training, and R&D are crucial elements in the performance of the aircraft sector. In a more dynamic view, Jacobs's externalities resulting from inter-industry knowledge spillovers are predominant, with a particular role played by higher education and the academic sector.

In the case of the *Communication* industry, the model outcome only statistically and significantly statistically considered the following variables: GD^k, *POP*, and *dummy* variables. The results indicate that all of the above variables have a positive

Table 5.1 MAR, Jacobs, and Porter externalities in high-tech industries

Industry	Aerospace	Communication	Computers	Biotechnology	Semiconductors	Lasers
Variables	Patents	Patents	Patents	Patents	Patents	Patents
GD^k	7.706***	7.609***	7.580***	5.593***	1.728	2.510**
	(2.539)	(1.302)	(1.104)	(1.278)	(1.203)	(1.065)
GD^p	0.188	−1.382	−1.629*	−2.246**	2.582***	1.264*
	(1.135)	(0.987)	(0.909)	(0.897)	(0.878)	(0.764)
HKIS	0.173	0.497	−0.0639	−0.231	0.970**	0.910
	(0.413)	(0.464)	(0.363)	(0.244)	(0.486)	(0.462)
EKC	3.94e-05**	−4.27e-06	1.49e-06	1.89e-05*	9.39e-06	−3.06e-06
	(1.56e-05)	(8.46e-06)	(7.03e-06)	(9.70e-06)	(9.50e-06)	(6.02e-06)
S	−0.146	−0.176	0.199*	0.0944	0.598***	0.0427
	(0.118)	(0.529)	(0.104)	(0.0861)	(0.174)	(0.0922)
CO	−0.000523	0.0173	−0.00511	0.129**	−0.0849	0.0141
	(0.0213)	(0.0799)	(0.00989)	(0.0612)	(0.0581)	(0.0513)
POP	−1.51e-07**	1.51e-07**	9.47e-08*	−1.88e-08	1.44e-08	−1.25e-07**
	(6.60e-08)	(6.09e-08)	(5.56e-08)	(6.77e-08)	(5.18e-08)	(4.88e-08)
RD	0.000761	−0.000844	−0.000444	0.000748	1.87e-05	−0.000409
	(0.000694)	(0.000605)	(0.000603)	(0.000602)	(0.000657)	(0.000493)
Dummy	−0.198	1.518***	1.573***	1.228***	0.799**	0.597***
	(0.291)	(0.310)	(0.260)	(0.346)	(0.348)	(0.222)
Constant	−1.969	−1.011	−0.771	−0.0699	−1.530**	−2.102***
	(1.420)	(0.827)	(0.555)	(0.510)	(0.671)	(0.511)
(2)	−1.166***	0.739***	0.521***	1.001***	0.486***	−0.763**
lnalpha	(0.358)	(0.120)	(0.112)	(0.126)	(0.133)	(0.334)
Observations	27	142	142	149	120	156

Standard errors in parentheses: $^{***}p < 0.01$, $^{**}p < 0.05$, $^{*}p < 0.1$

impact on the number of patents in an average region in the field of Communication. The production diversity GD^k turns out to be an important input for new knowledge creation in the Communication industry. According to model results, a 1 % increase in production diversity of an average EU region increases the number of patents by about 7,609 % (with 99 % significance). The findings suggest that this is so despite the fact that the telecommunications industry is reaching its phase of maturity. Internet technology and the diversification of products and services of its application continue creating new market opportunities for the EU communication industry. Thus a diversified local industrial structure is necessary for new Internet applications to generate new technological externalities.

Furthermore, the estimates presented in the table indicate that a 1 % increase in the number of population *POP* in an average region in a sample induces a 1.51e-07 % increase in the number of patents in Communication (with 95 % significance). The development of the Communication industry may depend on a more diverse industrial environment, and therefore innovative activity may depend on the concentration of diverse industries available in bigger cities.

The effect of a *dummy* variable shows a positive correlation between higher population density and number of patents. In fact, a 1 % increase in a dummy variable increases the number of patents in an average EU region in the communication field (with 99 % significance). This is in accordance with the results of other existing studies that have attempted to answer the question whether rapid progress and a decrease in the cost of communication and transportation over distance will attenuate agglomeration and will lead to more dispersion of economic activities. For example, Ota and Fujita ((1993) show that the development of information technology will lead to a greater concentration of front units in the city center and to a dispersion of back units in the far suburbs. Gaspar and Glaeser (1998) state that face-to-face interactions in the ICT sector are complementary, e.g. some face-to-face contacts will be replaced and conducted electronically, and the improved ICT may result in more face-to-face interactions and therefore strengthen agglomeration.

In summary, in the light of the literature on externalities, it could be suggested that diversified cities or Jacobs externalities benefit the EU communication industry's development more.

The results for the *Computer industry* indicate that the following variables; gdk, gdp, pop, s, dummy, and Constant lnalpha are statistically significant. All of the above variables accept that the innovation diversity indicator of a region GD^p has a positive influence on patenting activity in the Computer industry.

Computer applications have become an integral part of almost all industries, including their subsequent sectors, with the industries ranging from heavy manufacturing to agriculture. Industrial companies, such as those that produce goods, may use computers to create an automated time clock and earnings system. Thus, it is not surprising the new knowledge creation capacity of regions in the computer industry is being positively influenced by the region's diverse industrial structure GD^k (Table 5.1). Broadly, a 1 % increase in production diversity of an average EU region increases the number of patents by about 758 (with 99 %

significance). The computer-manufacturing industry experiences Jacobs's externalities from collocated software industry firms and other high-tech industries such as IT, electronics, aerospace, and semiconductors, in particular. In fact, findings for software industry experiences by Robert J. Kauffman and Ajay Kumar (2006) show that Jacobs's externalities in the software industry result from collocation with the semiconductor industry.

However, the estimated outcome suggests that the computer industry doesn't depend on the innovative diversity GD^p, i.e. innovative activity in other industries. The multitude of R&D projects of an average EU region pursuing many innovative goals across various high-tech fields will not stimulate new knowledge creation in the computer industry. The increase of the innovation diversity indicator by 1 % decreases the number of patents by approximately −1,629 (with 90 % significance). However, the computer industry may draw on the experience of other disciplines; the model result suggests the need to permanently adapt R&D projects to changing technological and market priorities in the same industry.

Furthermore, the results also indicate that the size of a region's population, production specialization, and population density are all statistically significant and positively influence patent activity in the EU computer industry. Thus, the computer industry appears to be highly concentrated and strongly determined by access to specialized production factors. For the computer industry, knowledge spillovers and specialized labor seem to be key assets. This suggests that as the computer industry gets closer to its technological maturity, its survival in the industry requires greater technological specialization and diversification of computer applications and operation/management in various fields. With many manufacturers producing a multitude of models, ranging from desktop computers to notebooks, a firm's major innovative activities should concentrate on specialized and advanced software applications. Hence, a positive and significant coefficient of specialization and production diversity measures would indicate that both increased specialization and diversity within a region are conductive to a greater number of patents in computer-related fields. This would support the possible existence of MAR and Jacobs's externalities in the computer industry across the EU regions.

In the case of the *Biotechnology* industry, the model outcome considered the following variables statistically significant: GD^k, GD^p, *EKC*, *COMP*, *dummy*. The results indicate that the dummy GD^k, *EKC*, and *COMP* have a positive influence on the number of patents, whereas GD^p is negative in an average region in the field of biotechnology.

Since the biotechnology industry is still in its early phase of growth, increasing competition in biotechnology has a significant positive influence on the acceleration of innovations. This would suggest the existence of Porter externalities. However, since production and innovation diversities determine patenting performance in biotechnology, these externalities could also suggest Jacobs's externalities. Except for the fact that Jacobs (1969) and Porter (1990, 1998) argue the opposite, Porter considers that competition is more conducive to knowledge externalities than a local monopoly would be. It should be emphasized that by local competition, Jacobs refers to the competition for new ideas embodied in economic agents.

An increased number of firms provide greater competition for new ideas, but greater competition across firms facilitates the entry of new firms specializing in some new product niche. This is because the necessary complementary inputs and services are likely to be available from small, specialist niche-oriented firms but not necessarily from large, vertically integrated producers. Most European companies specializing in biotechnology are small or medium-sized enterprises (SMEs). Therefore, rivalry is extremely intense. High research costs, the need to fully exploit patent protections before they expire, and the expenses of costly marketing induce close alliances and co-operation among biotech firms and R&D institutions. Therefore, both agglomeration economies (dummies) and the proximity to a qualified labor pool (ekc) as well as other biotechnology firms have a positive and significant impact on biotech patents. Other researchers have reached a similar conclusion. In fact, Prevenzer (1997) and Zucker et al.(1994) show that in biotechnology firms tend to cluster together in just a handful of locations.

Furthermore, the results of the model indicate that diversity across complementary economic activities is very conducive to innovation in the biotechnology industry. In fact, a 1 % increase of production diversity of an average EU region increases the number of patents by about 5.593 % (with 99 % significance), whereas, an increase in the innovation diversity indicator by 1 % decreases the number of patents by approximately −2,246 (with 95 % significance). The above results may simply mean pursuing many innovative goals across various high-tech fields will not stimulate new knowledge creation in the biotech industry. Although biotechnology is a very interdisciplinary science and draws on the experience of a number of different disciplines, the result of the model suggests the need to permanently adapt R&D efforts to changing technological and market priorities within the same industry.

Overall, the analysis suggests that biotechnology in the EU still remains a clustered economic activity and relies strongly on interaction with science-based university research. Biotech companies – especially those whose success depends on staying on top of new technologies and processes, increasingly want to be where new, hot ideas are percolating. This suggests patenting in biotechnology requires combining innovative specialization with industry diversity. The rate of innovation in biotechnology depends on the availability of a highly skilled and well-educated workforce.

In the case of the *Semiconductor* industry, the model results indicate that all the variables: *S, HTKS*, GD^p, and *dummy* have a positive impact on the number of patents. Other variables are statistically not significant. The assessment of the region's specialization indicators in high-tech industry *S* and high-tech knowledge intensive services *HTKS* both prove to have a positive and significant influence on patenting activity of an average EU region in the semiconductor industry. In fact, Table 5.1 indicates that an increase of regional production specialization in the high-tech industry by 1 % increases the number of patents in semiconductors by 0.598 (with 99 % significance). These findings, along with the positive and significant influence of the population density dummy on patent activity, suggest that the semiconductor industry in an average EU region is characterized by MAR externalities. The increased concentration of the semiconductor industry within a specific geographic area facilitates knowledge spillovers and innovation processes. The study results are in accordance with the overall outcome of Ketelhöhn's (2006) survey on the evolution of the geographic concentration of the US semiconductor industry from the early 1960s to 1997 and sources of knowledge spillovers for the industry. The author found that clustering and co-location with important buyers is associated with higher levels of industry innovation. The presence of MAR externalities in the EU semiconductor industry is in line with the mainstream literature on dynamic externalities. Mature industries are associated with MAR externalities as the industries focus on cost-saving process innovations and therefore benefit from specialization.

For the second variable, a 1 % increase in production specialization in high-tech knowledge-intensive services, *HKIS*, of an average EU region, increases the number of patents by about 0.97 % (with 95 % significance). The availability of knowledge-intensive high-tech services may also be associated with the industry's experience with outsourcing and offshoring. The latter two phenomena have become even more popular in the case of mature industries, where cost reduction is a crucial determinant for competitiveness. However, there is insufficient empirical evidence of the scope or scale of potential outsourcing and offshoring in the EU. Some relative evidence of offshoring in a sample of US semiconductor firms was delivered by Leiblein and Reuer (2004). The empirical evidence on outsourcing and offshoring in the EU semiconductor industry in relationship to technological maturity, location proximity, and externalities can help to better explain the role of space in the process of knowledge creation and regional growth.

The semiconductor industry is one of the most research-intensive high-tech fields. Its R&D results in micro and nano electronics strongly influence innovation processes in other fields, such as communication or IT. It is not surprising, therefore, that the results of the current model show that an increase of the innovation diversity indicator GD^p by 1 % increases the number of patents by approximately 2,582 (with 99 % significance). A number of researchers have provided evidence for the learning-by-doing spillovers, which characterize the semiconductor industry (Irwin and Klenow 1994; Stokey 1988; Lucas 1993). According to evidence provided by Lucas (1993), the importance of technological advancement in the semiconductor industry for the performance of other high-tech industries causes diverse learning spillovers. Furthermore, as the semiconductor industry enters its decline stage, innovation diversity and learning-by-doing spillovers may also improve product quality and contribute to potentially new innovations (Bailey and Huang 2010; Malerba and Orsenigo 1994).

Finally, in the case of the *Laser* technology industry, the model outcome GD^k, GD^p, *POP* and the *dummy* variables are considered to have a statistically significant and positive effect on the patent activity of an average EU region in the laser technology field.

Lasers are a key element of research across all domains of industry and science, from life science to material research. From supermarket barcode scanners to communication highways, from less invasive surgery to automobile production – the special properties of laser light are invaluable. Therefore, it is not surprising that both production and innovation diversity indicators positively influence and significantly impact patent activity in the laser technology field. A 1 % increase in the production diversity and innovation diversity indicator causes a 251 % and 1,264 % increase in laser technology patents, respectively (with 95 % and 90 % levels of significance).

Furthermore, Table 5.1 emphasizes the positive and significant impact of the population size *POP* and dummy variables for the average EU region's patent activity in laser technology. This may signify that the laser technology industry is located mainly in densely populated regions that are home to a university with varied S&T departments. Similar evidence has been supplied by Fritsch and Medrano (2010), who argue that the early stage of laser knowledge adoption is especially prevalent in large agglomerations. Furthermore, the region's high absorptive capacity and existing tacit knowledge are crucial for commencing, conducting, and diffusing research findings in a new laser technology field.

As the laser technologies industry shows characteristics of an industry still in the early phases of evolution, the heterogeneous and diverse industry structure allows for inter-industry knowledge spillovers and successful adoption of innovations, and therefore shows the predominance of Jacobs's externalities.

To summarize this section: despite attempts to cover many variables affecting dynamic externalities and patenting activity in the high-tech sector in the EU regions, the above models are not without critics. The value of the above model estimations is limited by the fact that the use of input–output tables only measures market-based transactions and overlooks the critical contribution made by soft factors such as the region's socio-economic structure, companies' size, and the process of industrial modernization. It is clear that the econometric models are unable to take into account all of the industry dynamics. The purpose of this study was to answer the question concerning the extent to which geographical proximity, industrial structure, and technological maturity influence local knowledge spillovers (whether intra- or inter-industry spillovers) and innovations of high-tech regions.

5.5 Summary and Conclusions

After more than two decades of investigating the importance of MAR, Jacobs' and, more recently, Porter's externalities, there is no and probably never will be any consensus in the literature concerning the role of these externalities in localized knowledge flows and the regional innovation process. This is because every industry has different agglomeration needs during different phases of its life cycle. To test this, the above chapter presented an econometric framework that allowed us to assess the role of agglomeration externalities in patent performance in six types of high-tech industries for the samples of EU regions. In addition, the estimates findings were discussed with reference to the technological maturity of the above industries.

Agglomeration externalities in the sense of Jacobs's externalities resulting from inter-industry knowledge spillovers turned out to be predominant in the aerospace, communication, biotechnology, and laser instruments industries; however, this was all for different reasons. The importance of Jacobs' externalities for the aerospace industry is explained by the fact that the industry is still in its growing stage of technological maturity and its development convergences with the expansion of other industries (especially IT), and therefore requires close proximity to institutions of higher education. The same kind of reasoning can be applied to the role of Jacobs' externalities in the biotech industry, which is still in its early growth phase. The rate of innovation in biotechnology depends on the strong interaction with science-based university research and on the presence of other industries, such as pharmaceutical, chemical, health care, food, etc. Therefore, the innovative activity of biotechnology firms is determined by a combination of both innovative specialization, industry diversity, and competition externalities (Jacobs, MAR, Porter). In the same way, the early technological maturity of the laser technologies industry suggests that a heterogeneous and diverse industry structure will benefit innovations in the laser and optic fields more (Jacobs' externalities). In contrast, the communication industry has reached its phase of maturity, and thus would benefit more from the diverse industrial environment available in bigger cities.

As far as the semiconductor industry is concerned, the estimates findings indicate the presence of MAR externalities, which is in line with the mainstream literature on dynamic externalities. The industry enters its decline stage in which specialization and learning-by-doing spillovers may improve quality and be more cost efficient. The importance of cost reduction for the semiconductor industry may also explain the crucial role of high-tech knowledge-intensive services for its innovative potential, which, on the other hand, may be related to the industry's experience with outsourcing and offshoring. This evidence opens the door for further research and exploration of the relationship between outsourcing and offshoring activities and technological maturity.

In general, the findings suggest that industrial diversity and agglomerative density have higher impacts on innovation and technological change in the EU regions than industrial specialization. Nevertheless, diversification efforts should focus not only on industrial activity but also promote the development of the high-tech services sector and access quality institutions of higher education.

Chapter 6
Knowledge Spillovers, Technological Change, and Regional Growth. Empirical Evidence of the EU Regions

6.1 Introduction

Regional disparities can no longer be defined only in terms of statistical differences in values of standard macroeconomic indicators. Knowledge and technological advancement matter more and more in defining both the level and growth rate of a given region's GDP. The following chapter aims to analyze the impact of knowledge spillovers and technological change on economic growth and the catching-up process of the EU regions.

The exact impact of technological change and knowledge spillovers on regional growth is very difficult to measure, just as it is difficult to measure the process of innovation creation with a simple linear model. The empirical evidence of the role of innovations, technological change, and economic growth suggests there are also many feedback loops and feedbacks of activity across the processes of innovation, technological change, and GDP growth.

As explained in Chap. 1, changes in productivity and moving up the value chain were emphasized by growth theories as an important measure of technological change (Solow 1956; Arrow 1962; Romer 1986). Productivity rises through new innovations and technological advancement. However, technological change can happen only if every worker improves his or her skills by undergoing a learning process. Furthermore, a highly educated and multi-skilled workforce generates spillover effects, which increase productivity and create innovations that may impact the growth of knowledge-intensive and high-tech industries. On the other hand, increasing specialization in the production of high-tech goods and services creates even greater demand for high-skilled labor and advanced technologies, which further contributes to productivity.

As explained in Chaps. 4 and 5, technological advancement in one industry increases the productivity of innovating firms belonging to this industry and other industries through dynamic externalities and knowledge spillovers; therefore, it influences the innovative performance of other regions and locations.

M. Runiewicz-Wardyn, *Knowledge Flows, Technological Change and Regional Growth in the European Union*, Contributions to Economics,
DOI 10.1007/978-3-319-00342-9_6, © Springer International Publishing Switzerland 2013

With regard to the preliminary literature discussed in Chap. 1, the following chapter takes an approach in which the region's technological catching up and GDP per capita growth depend on three main factors: innovative intensity, human capital, and technological change. The first two factors are assumed to shape the dynamics of both the region's absorptive capacity and the efficiency of its "innovation systems," whereas the last one refers to a region's ability to undertake technological change and move up the value chain. Therefore, the model is rooted in the technology gap tradition and is based on the theoretical literature of Kaldor (1960), Cornwall (1976), Griliches (1979), Abramowitz (1979, 1986), and Romer (1990). The common idea of the above theories is the positive relation between technology and economic growth. The following book attempts to complement the relevant literature by investigating the experience of enlarged EU regions, and the role in these regions of lifelong learning, innovation propensity, high-tech services (that could proxy the potential capacities for outsourcing activities), and technology proximity in technological catching up and GDP per capita growth.

The role of innovations and technological catching up has become more pervasive because in technologically leading regions, as well as in many lagging regions, more than 40 % of all structural funds are currently related to research and innovation. All regions in Europe rely on innovation expenditures for peripheral regions to catch up and technological leaders to remain ahead (Wintjes and Hollanders 2011). Therefore, it is important, both from the policy and theoretical perspective, to discuss the relationship between knowledge, innovation, technological change, and regional growth, taking into account the EU regional landscape.

This chapter is structured as follows: the first two sections focus on a survey of closely related studies of the relationship between technological change, knowledge spillovers, and catching up of EU regions. The third section empirically tests the correlation between innovation propensities, stock of knowledge, and catching up of EU regions. The fourth section includes the construction of the dataset, the statistical model used in the analysis, and provides a thorough description of the model variables. The fifth section provides the main findings on the applied model framework in the analysis of technological change and catching up in EU regions. The last section discusses employment trends in the high-tech sector and examines their relation to economy-wide fluctuations. The chapter ends with a summary and the main conclusions.

6.2 The Relationship Between Technological Change, Knowledge Spillovers, and Catching Up of EU Member States Regions: Empirical Studies Literature Review

6.2.1 *Technological Gap and Economic Convergence of Regions*

In the spirit of the endogenous growth and "catching up" literature, knowledge and technological advancement are primary elements of economic growth models. The "stock of knowledge" is available to the economy as a result of education, learning by doing, and R&D efforts. The ability to upgrade the existing "knowledge stock" and absorb the latest technologies determines the level of productivity and proxies technological change (Solow 1956; Arrow 1962; Romer 1986; Autor et al. 1997; Berman et al. 1994; Murphy et al. 1998). The level of productivity is assumed to reflect the degree of technological sophistication and is responsible for the level of technology gap between countries or regions. If the level of productivity is assisted by an increasing share of knowledge-intensive employment, one could expect a region's per capita income level to increase.

Despite theoretical differences, many contributions to the empirical literature on regional productivity and growth performance share a common empirical framework. Mainly, they give evidence that innovation, technology gap, and knowledge spillovers are possible sources of income convergence in the US and EU landscape (Verspageri 1993; Amable 1993; Howitt and Mayer-Foulkes 2005; Stokke 2008).

Verspageri (1993) measured the relationship between the technology gap and cumulative growth on a sample of 114 countries between 1960 and 1985. He proxied the initial technology gap by GDP per capita. The results show that technology gap dynamics determine a country's knowledge stock and innovative activity. A similar approach was applied by Amable (1993), who examined whether catching up took place during the period 1960–1985 in 59 developed and developing countries. He used the percentage of the US level of real GDP per worker as a proxy for the initial technology gap. His system of equations can be summarized as follows: the labor productivity growth rate is assumed to depend on the initial technology gap, the rate of equipment investment, the percentage of the concerned age group engaged in primary education, and the ratio of real government expenditures (less defence and education) to real GDP. Equipment investment, innovative activity, and the level of schooling are endogenized.[1] Amable's results are positive for the role of primary education and government expenditure, and catching up in real GDP per worker.

[1] While the reference to Amable (1993) is fundamental for the literature review in this chapter, it should be noted that Amable's simultaneous system of equations fails to satisfy the order condition for identification.

Targetti and Foti (1997) found that the process of catching up leads to growth in productivity and an increase in exports competitiveness through the technological spillover effect. They investigated the interactions between productivity, demand, and exports, based on the approaches of "cumulative causality" of Kaldor and the "technological gap" on a sample of 25 countries over a period of 38 years. The initial technology gap was proxied by a log of the GDP per worker ratio between the leader and the followers.

Although an increasing number of empirical studies have investigated issues of economic convergence in the EU, the impact of technological change in regional convergence has so far received less attention, especially when the latest enlargement is considered.

Estimating the model based on the data for 267 NUTS-2 regions of the EU27 during the period 1995–2006 that was used by Alexiadis et al. (2010), yielded some interesting results. They provided some evidence that EU regions with a low productivity level grow faster than high-productivity areas. However, given the extremely slow convergence rate, it would take a very long time for all the regions to reach a common level of productivity. More important output of their research is that regions with an initial low level of innovation exhibit relatively higher growth rates. Still, the above authors concluded that a high technological gap does not necessarily imply a technologically lagging region's ability to adopt technology – a large gap may constitute an obstacle to convergence.

Two other authors, Alexiadis and Tomkins (2008), came to similar conclusions by applying a different approach. They attempted to address the impact of adoption of technology on labor productivity and GDP per capita convergence. They used data for the 258 NUTS2 regions of the EU26 over the period 1995–2004. The results suggest that the EU regions exhibit a tendency towards convergence in terms of both productivity and per capita output; however, convergence appears to occur faster in terms of labor productivity. The overall findings of their study suggest that adoption of technology – although it might be the best 'vehicle' for lagging regions to converge with leading regions – is a process which may be difficult for lagging regions, especially during the early stages of development when conditions are least supportive.

Another effect of growth through the inflow of superior technologies from abroad was evidenced by Rapacki and Prochniak (2009), who show that enlargement significantly contributes to economic growth and income-level convergence between the EU-10 countries and the EU-15 ones. The former grew faster, on average, than the latter between 1996 and 2007, even though their initial income level was much lower. Furthermore, the major driver of this growth was gross capital formation and gross fixed capital formation, both of which revealed positive and significant correlations with the GDP growth rate. Similar evidence of the positive role of FDI in international technology transfer was found within a limited number of NUTS2 regions in the EU-10, as examined by Tondl and Vuksic (2003).

In terms of the findings on the EU-15 countries, Fagerberg et al. (1997) tested the link between technology, growth, and unemployment using data for 64 EU regions in the 1980s. The growth of GDP per capita and the employment rate were found to be clearly interdependent: higher growth is likely to lead to the creation of new jobs, and vice versa. Nevertheless, if the additional labor turns out to be rather unproductive, adding it to the population may actually lead to a fall in GDP per capita (Costa and Kahn 2000). To some extent, a similar outcome was achieved by Greunz (2003) in her study on the role of the technology gap in catching up at a regional level based on a sample of 153 European regions in the period 1989–1996. She showed that, on average, the growth rate of GDP per capita is assumed to be a positive function of the rate of growth of productivity and structure of the productive system; therefore, the majority of EU regions evolve towards higher steady state ratios of relative GDP per capita than their current ones.

Furthermore, several studies on EU regional growth in different periods of analysis show that the Kaldor-Verdoorn law continues to be strong and robust in the EU regions' growth experience. Fingleton and McCombie (1998), Pons-Novell and Viladecans-Marsal (1999) found evidence of substantially localized increasing returns to scale in EU manufacturing at the NUTS2 and NUTS1 level, respectively. These findings once again provide support for the endogenous growth theory, and they prove that manufacturing is the engine of economic growth (Kaldor's first law), and overall productivity growth is positively related to manufacturing output growth (Kaldor's third law).

The literature shows there is a positive relationship between specialization in the knowledge-intensive sectors (high-tech manufacturing and knowledge-intensive services) and a region's per capita income level. Mora (2008) found that the higher the economic structural importance of high-tech sectors in a region is, the higher the level of development will be (e.g. employment and income). Moreover, these sectors have played an active role in opening new markets and technological opportunities via the globalization processes in the EU (Cooke 2005; Narula 2004; Patel and Vega 1999). Finally, in the "new economic geography" models, increased high-tech specialization (as a result of the reduction in transport costs and of increased integration) leads to the generation of regional differences in growth. Ledesma (1999) investigated interactions between export-led growth, productivity gap, and innovative activity on a sample of 17 OECD countries over a 29-year period. The initial technology gap was proxied by the level of labor productivity relative to the US. Ledesma's findings show that the process of catching up leads to growth in productivity and to an increase in high-valued exports (as a result of the diffusion of technology).

Another example is the study of Cortright and Mayer (2001), who investigated the link between growth and "high-tech" specialization (measured as employment concentration, patent activity, and venture capital flows) in US metropolitan areas. They found that high-technology employment is concentrated in only a few industry segments in the majority of regions, and that it has triggered their phenomenal growth. Saxenian (1994), and Markusen (1994) reached similar conclusions in their studies. Some evidence was found for the role of particular high-tech sectors in a region's productivity and income catching up. For example, the findings of Piatkowski and van Ark (2004), Jorgenson and Stiroh (2000), and Jorgenson (2003) show that the ICT sector plays a significant role in closing the productivity gaps in Europe and the US. Meanwhile, empirical evidence by Jorgenson et al. (2008) has pointed to a fundamental shift in the production cycle of semiconductors, involving the more rapid introduction of faster and better chips, which has spurred massive investment and, therefore, regional economic growth.

Moreover, most of the empirical evidence literature emphasizes the role of agglomerations externalities in knowledge spillovers, technology diffusion, knowledge-intensive specialization, and regional growth (Cappellen et al. 2003; Giannetti 2002; Paci and Pigliaru 2002). For example, Giannetti (2002) tested the causal relationship between integration and convergence in the EU by assuming that regions specialized in the high-tech sectors will benefit more from increased knowledge spillovers while empirically considering the 108 NUTS2 regions in 11 countries – the EU-15, excluding Austria, Finland, Sweden, and Luxembourg. Knowledge spillovers bring convergence across regions that can competitively produce goods involving high-end technologies. As the high-tech sector contributes more to total output, countries also converge in terms of per capita income. However, within countries exist some regions that are devoted to traditional goods; hence, disparities are amplified across regions within a country.

Summing up, a huge number of empirical studies has addressed regions' catching-up issue in terms of income, productivity, and, more recently, technological capabilities. Broadly speaking, their findings confirm that domestic technological capabilities – in terms of R&D activities, infrastructure, and human resources – are key factors in the enhancement of technological catching up and economic growth processes. The difficulties in coming to definitive conclusions based on these studies arise from the fact that their results are based on the experience of advanced or successful economies, such as the US or EU-15. Considering that the EU regions' growth is determined by a continuous process of integration, the role of technology transfer and knowledge spillovers should be monitored further and tested against the existing literature.

6.2.2 *Absorptive Capacity, Innovativeness, and Catching Up of Regions*

More specifically, the framework of the endogenous growth models and empirical evidence argue that two dimensions determine a region's ability to narrow its technological gap. The first is its absorptive capacity, e.g. its ability to imitate foreign advanced technologies. The second is its innovative capability, namely the extent to which it is able to produce new, advanced knowledge. The narrowing knowledge absorption and innovation gaps between regions improve a region's productivity level and move it up the value chain towards specialization in knowledge-intensive and high value-added activities. As shown in Chap. 1, economic theory has argued strongly that education and learning should be considered important factors for narrowing the knowledge absorption and innovation gaps (Lucas 1988; Vanhoudt et al. 2000; Tondl 2001; Badinger and Tondl 2003). In addition, a higher level of education is considered to have a direct impact on research capacity (Romer 1990).

Only a few empirical studies have addressed the problem of technological change and the role of education for economic growth. Most of them concern issues of the experience of the US (Gottlieb and Fogarty 2003; Glaeser et al. 1995; Benhabib and Spiegel 1994; Eaton and Kortum 1996), and, to lesser extent, the EU regions (Badinger and Tondl 2003; Tondl 2001; Gottlieb and Fogarty 2003; Glaeser et al. 1995; Benhabib and Spiegel 1994; Tondl and Vuksic 2003; Fidrmuc 2000; Campos 2002).

Gottlieb and Fogarty (2003) studied 75 large US metropolitan areas and concluded that the higher number of residents with at least a bachelor's degree in 1980 was associated with higher per capita income and employment levels 17 years later. Because of knowledge spillovers, the most educated metropolitan areas in the US had their per capita incomes 20 % above the average, whereas the least educated had incomes 12 % below the average. The authors defined "education attainment" as the proportion of adults aged 25 and older with 4 or more years of college. The average educational attainment amounted to 25 % for the top metropolitan areas (Atlanta, Georgia Austin, Texas, and Boston, Massachusetts).

Glaeser et al. (1995) argue that human capital is significant to explain the growth of cities. Specifically, education levels are closely related to subsequent income and population growth. Simon (1998) found "a positive, large, and persistent relationship" between the level of human capital and employment growth at the metropolitan level. Meanwhile, the studies of Eaton and Kortum (1996) provide evidence that human capital has a positive and statistically significant effect when interacted with the technology gap. In fact, as Nelson (2004) emphasizes, because of the increasing complexity of modern technologies, advanced research and training has become a prerequisite for the ability not only to develop, but also to understand and assimilate technology.

Furthermore, empirical studies have found that education and technological progress are of much higher importance than factor accumulation, based on the experience of OECD and EU regions (Bassanini and Scarpetta 2002 for the OECD and Vanhoudt et al. 2000, and Fagerberg et al. 1997 for the EU regions).

An endogenous growth model, such as the one used by Badinger and Tondl (2002), investigated EU regions in the 1990s and linked human capital explanations to the catching-up theory. Their results indicate that regional income growth is positively linked to labor participation and its educational attainment. Higher education is also an important prerequisite for lagging regions for technological catching up. One of the aforementioned authors showed in his previous study that income and productivity of the Southern EU regions are positively correlated with school enrolment (Tondl 2001). The author was in line with the findings of Vanhoudt et al. (2000) on a positive elasticity of human capital, measured as educational attainment for EU regions of 0.18 (when estimating a production function for 1996).

However, they provide slightly mixed evidence for growth regressions of EU-10 countries. Tondl and Vuksic (2003) suggest the rate of secondary school attainment in the population plays an important role in regional growth. This should not be surprising, since the EU-10 shows a fairly high level of educational attainment. Fidrmuc (2000) also proposes that secondary education has a positive impact on growth, whereas Campos (2002) argues that it has a negative impact. Campos suggests that in many Central and Eastern European countries, despite the high level of educational attainment of the average worker, the composition of the stock of human capital (in terms of their work occupations) has proved to be inadequate in addressing the needs of a modern market economy.

With reference to innovative capability – according to the new growth theory – productivity growth can be generated either by home regions' R&D or derived from technology transfer and knowledge spillovers from other regions, e.g. by innovating or imitating technology produced elsewhere. The step that specifically links the performance of RIS to productivity growth has been relatively neglected in the literature. An obvious difficulty lies in the simultaneity of productivity and innovation performance at regional level, together with the lack of comprehensive data. Usually, studies focus on firm-level analyses, where data is available on different types of innovation (product and process) introduced by firms. The available empirical evidence generally suggests that the direct effect of research activity or innovative capacity on productivity growth – frequently measured by the ratio of R&D expenditures to output – is positive and substantial. An early example is the firm- and industry-level work of Griliches (1980) and Jaffe (1989) for the United States.

In his research framework, Jaffe (1989) observes significant positive effects of R&D spillovers on the firm's position in the technological space. The results of Coe and Helpman (1995) and Guellec and van Pottelsberghe de la Potterie (2004) reveal the important relationship between R&D expenditures and productivity growth for the OECD countries. Griffith et al. (2004), and Cameron et al. (2005) find indirect and direct effects of R&D on total factor productivity by facilitating technology transfer. Griffith et al. (2004) used panel data on manufacturing industries across 12 OECD countries from 1974 to 1990, while Cameron et al. (2005) focused on a panel of UK manufacturing industries observed between 1971 and 1992. The above study also shows the positive influence of knowledge spillovers coming from abroad (by import-weighted R&D activity of a country's trading partners) on domestic productivity growth. Similar findings were achieved by Khan (2006), who focused on French manufacturing industries and concluded that R&D affects total factor productivity growth primarily through the direct innovation channel, allowing spillovers within manufacturing industries across countries.

Vogel (2012) investigated two channels through which R&D and human capital may affect regional total factor productivity growth in the manufacturing sector, using panel data on 159 EU-15 regions from 1992 to 2005. Relying on endogenous growth model assumptions, she allowed R&D and human capital to influence productivity. Furthermore, the model allows for conditional convergence to a long-run level of total factor productivity relative to the frontier. Her results suggest that total factor productivity convergence takes place over the selected sample period and that geographic distance from the technology frontier matters. Similar findings, based on patents as an imperfect measure for innovation, were achieved by Bosco and Brugnoli (2010). The authors tested the relationship between RIS and the productivity-based knowledge production function approach, covering 29 OECD countries, with 323 regions for 1995–2008. They found that innovation inputs and outputs are positively related to productivity. Castellani and Pieri (2011) address this issue by investigating the extent to which R&D offshoring is associated with the productivity dynamics of European (NUTS2) regions. Particularly, they explore whether R&D investments abroad have a different impact than those in manufacturing or other business activities. The authors claim that offshoring regions have higher productivity growth, but this positive effect fades with the number of investment projects carried out abroad. Still, a large and positive correlation emerges between the extent of R&D offshoring and home region productivity growth, supporting the idea that carrying out R&D abroad strengthens European competitiveness.

Other evidences show that investment, whether from tangible capital or human or R&D expenditures, plays a central role in the level and dynamics of productivity. For example, Morrison and Siegel (1997) included R&D investment, high-tech capital investment, and human capital investment in a single analysis, and found all of them to be significant determinants of productivity growth in US manufacturing, with R&D having the strongest impact. Future empirical research should include many broad types of investment and capital in any productivity analysis. Only by considering the quantity and quality of all inputs, can one correctly estimate the marginal importance of each type of investment. The evidence provided by Greunz (2003), in a study carried out on a sample of 153 EU-15 regions on a NUTS2 regional level for the period 1989–1996, pointed out the importance of capital-embodied technology transfer and manufacturing investment for productivity growth in the EU regions.

In addition, many empirical studies have found a positive relationship between product innovation and employment at the firm level (Greenan and Guellec 2001; Harrison et al. 2008). Positive empirical evidence is found mostly in high-income and developed countries (Pianta 2005; Djellal and Gallouj 2007). Greenan and Guellec (2001), and Harrison et al. (2008) found the relationship to be positive, whereas Van Reenen (1997) found it to be weakly positive or of no significance. Furthermore, Evangelista and Savona (2003) found this relationship to be negative. These results are sensitive to the method and data used and the type of economy analyzed.

Summing up, although the importance of absorptive capacity and innovative ability for the growth process is widely acknowledged in modelling exercises, the empirical literature has not yet achieved a systematic understanding of how these two dimensions evolve and influence the technological catching-up process in the enlarged EU-25 regions. Sections 3 and 4 of the following chapter will take further steps in an attempt to explore how regions' absorptive capacities and innovative activities relate to changes in the technological gap between the EU regions and GDP growth.

6.3 Knowledge Spillovers, Agglomerations, and Regional Growth in Empirical Studies Literature

There is a number of empirical studies confirming the presence of a positive spatial autocorrelation (clustering) process in innovative activity in Europe (Rey and Montouri 1999; Lopez-Bazo et al. 1999; Le Gallo et al. 2003; Moreno et al. 2003; Abreu et al. 2004; Fritsch and Slavtchev 2007; Das 2009). Hence, it emphasizes the importance of allowing for spatial dependence in productivity growth, specialization, and regional catching-up models.

Notably, Rey and Montouri (1999), Lopez-Bazo et al. (1999), Le Gallo et al. (2003), and more lately Abreu et al. (2004) have studied the effects of spatial autocorrelation associated with changes in income, innovation, and productivity. In most cases, the authors have found strong patterns of global and local spatial autocorrelation, with some evidence that temporal changes in spatial autocorrelation are associated with changes in regional income dispersion. Their results indicate that being located close to other regions with high technology growth rates results in higher domestic technology growth (*ceteris paribus*). Consequently, these regions that are further behind the technology leader experience higher total factor productivity growth, due to their high social capability. The conclusion of the authors' results suggests the possibility that technology levels are converging locally and that countries would benefit from closer ties to the local technology leader.

Keller's (2001) work shows that the link between R&D spending in one industry and productivity in another can be used to estimate knowledge spillovers. His analysis is based on data for two- and three-digit manufacturing industries in Canada, France, Germany, Italy, Japan, the UK, and the US during the years from 1970 to 1995, and it gives substantial evidence that positive knowledge spillovers affect only geographically proximate firms.

The results of the study run by Das (2009) confirm the influential role of corporate R&D on the spatial distribution of innovation in Norway. The study indicates the strong existence of spatial autocorrelation between regions, suggesting spillovers of externalities between Norwegian labor market regions. Moreover, the study distinguishes between positive spatial autocorrelation (which occurs when high or low values of the variable tend to cluster in space) and negative spatial autocorrelation (which occurs when a geographical area is surrounded by neighbors with very dissimilar values).

A study by Moreno et al. (2003) on EU regions provides evidence that patenting activity in a certain region tends to be correlated to innovation performed in neighboring areas. Moreover, local indicators show the existence of significant local clusters of highly innovative regions in the EU, e.g. in West Germany. A spatial association is also found at the sectoral level, pointing to the formation of specialized clusters in different sectors in innovative regions.

Empirically, several channels of knowledge flows that influence productivity should be considered. These are usually derived from inter-industry sales (van Meijl 1995), investment flows (Sveikauskas 1981), or from a full input–output framework (Terleckyj 1982; Wolf and Nadiri 1993; Sakurai et al. 1996). In contrast, weighting by technological distance measures accounts for the fact that the absorption of knowledge spillovers is mediated by the technological proximity between receiver and transmitter. Such distance may be measured by the type of R&D performed (Goto and Suzuki 1989), the qualifications of researchers (Adams 1990), the distribution of patents between patent classes (Jaffe 1986) or patent classifications and citations (Verspagen 1997). In a sense, technology flow matrices combine the two concepts of technological and 'market' proximity by identifying originators and (potential) users of a technology or an innovation. Scherer's user-producer matrix, as well as the Yale matrix, was derived from patent statistics (Scherer 1982; Putnam and Evenson 1994).[5]

All the studies indicate the existence of spatial autocorrelation in innovation and technology levels in neighboring regions, suggesting spillovers of externalities. This can be observed in studies by Cheshire and Magrini (2000), who analyze inter-regional migration flows; by Paci and Pigliaru (2001), who analyze productivity spillovers across EU regions; and Funke and Niebuhr (2001), who analyze regional R&D spillovers. Nevertheless, since we wish to examine the total spillover effects between regions, rather than single effects, for which data is not always available, we shall focus on growth spillovers. This effect will be analyzed by using a spatial econometric model, as suggested in the spatial econometric literature (Anselin 1988; Anselin and Florax 1995; Kelejian and Prucha 1998).

Traistaru et al. (2003) found a positive relationship between knowledge spillovers proxied with a measure of FDI intensity and regional growth, as well as between regional manufacturing specialization and regional growth. Her results show that high growth rates are associated with high initial levels of GDP per capita. The change in regional specialization is also positively related to regional growth. Giannetti (2002, p. 539–567) investigates the relationship between regional specialization, investment-trade flows, and technology transfer more precisely. He argues that international knowledge spillovers affect only certain high-tech sectors. First of all, integration and greater exchange of knowledge among EU regions spur growth and bring convergence among regions. However, they also create harsher disparities within individual countries.

The main purpose of Ejermo and Grasjo (2008) was to establish to what extent proximity to R&D and university-educated labor could explain regional export specialization. This was achieved by estimating knowledge production functions, with total export value and number of high-valued exports in Swedish municipalities from 1997 to 1999 as outputs. The results show that accessibility to university-educated labor and company R&D, both at intra- and inter-regional level, have a positive impact on the value of export performances.

To sum up, the transition towards a knowledge economy should be considered as a relationship between innovations and technological changes in a spatial structure. Deriving the economic impact from technological advancement and innovation depends on dynamic interactive processes involving individuals, firms, and institutions, which absorb, apply, and diffuse technology across space. Empirical studies, using spatial dimensions of innovation processes in explaining productivity and income growth dynamics, have only recently started to emerge. Many empirical studies have confirmed a relatively high spatial autocorrelation in innovation and technology levels in neighboring regions, suggesting that knowledge spillovers are related to productivity growth. The results of these studies suggest that technology levels are converging locally and that regions would benefit from closer ties to the local technology leader. However, the results depend to some extent on the construction of the technological spillovers variable and a spatial dimension of analysis.

6.4 Do R&D Expenditure and Human Capital Promote Productivity and Growth in EU-25 Member States Regions?

The aforementioned literature has inspired the author to test what the level is of the spatial autocorrelation of human capital, innovation, and productivity growth in the case of the EU regions. Following theories created by Solow (1956), Arrow (1962), and Romer (1986), in which changes in productivity have been emphasized as an important measure of technological change, there should be a positive correlation between innovation propensity and productivity growth of the EU regions. The following section aims to review the average dynamics of the EU regions.

In general, according to the endogenous model, greater innovativeness indicates that higher overall productivity should be expected. However, the presumption that R&D and innovation are the key drivers of economic growth is difficult to reconcile with the empirical evidence for the EU regions. Figure 6.1 shows that productivity growth may not apparently be correlated with innovation propensity growth (as it is proxied by total intramural and extramural R&D expenditures on EU NUTS2 levels). One reason for this negative relationship is that R&D intensity is assumed to have the largest impact on productivity in high-tech industries and services. In particular, basic research is regarded as an essential component for a firm operating in a high-tech industry. Besides, if high-tech companies are characterized by investing a larger share of their budgets into R&D activities, they may also be expected to generate greater value added than firms active in low-tech sectors.

Another explanation may be that in many EU regions with a dominant share of public research institutions, R&D has been slowly incorporated in new technology

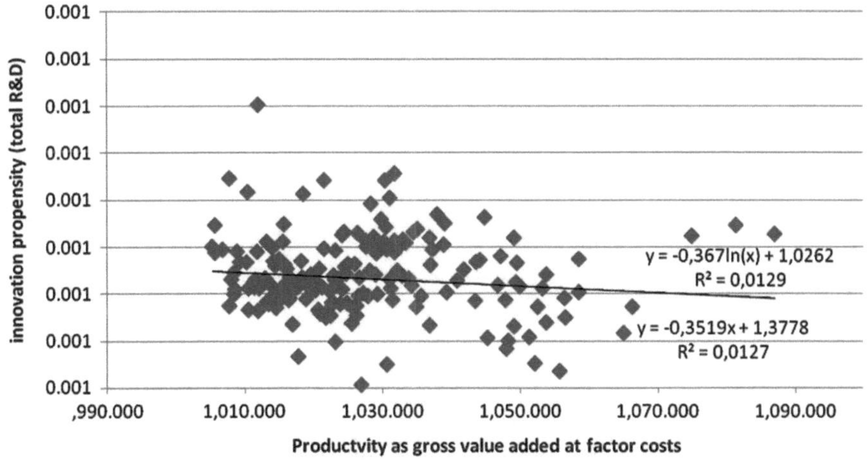

Fig. 6.1 Average dynamics of innovation propensity (total R&D expenditure) and productivity growth (rate of regional gross value added (GVA)) at basic prices in EU regions (1998–2008) (Source: own calculations)

areas, especially ICT, which has spurred productivity growth in recent years (Lichtenberg 1993; Guellec and van Pottelsberghe de la Potterie 2004). Much of the public R&D addresses the problems of health, environment, and defence and, therefore, has no direct impact on productivity, whereas university R&D is more concerned with basic research and may eventually have an impact on technological innovation at a later stage (Guellec and van Pottelsberghe de la Potterie 2004). This lack of flexibility of publicly funded R&D has contributed, at least partially, to the negative impact of R&D expenditure on productivity.

Business R&D is very important for a company's productivity growth (Temple 2002). Nevertheless, at the EU aggregate level, the business R&D expenditure level has remained almost unchanged since 2000 and is below 1 % of GDP in barely three quarters of NUTS2 regions (Eurostat 2010). The low return on R&D may also explain the small potential of R&D investments to increase productivity.

Moreover, the process of implementation of R&D-based technological innovations, aiming to improve the productivity of each worker, should be accompanied by organizational and learning processes. Some industries have higher rates of innovation and/or involve more complex technologies; therefore, they may require greater investment in learning in addition to R&D activity. The negative relationship between innovation propensity and productivity growth in the EU regions may also stem from inadequate investments in the learning and traineeship process.

Moreover, if R&D activity is to be associated with the process of searching for better technologies during the actual production process or use of technology, then it can be closely linked to productivity improvements as a result of "learning by doing"; at the industry level, this means the sum of the experiences of all firms comprising that industry. The "learning by doing" takes place if each company

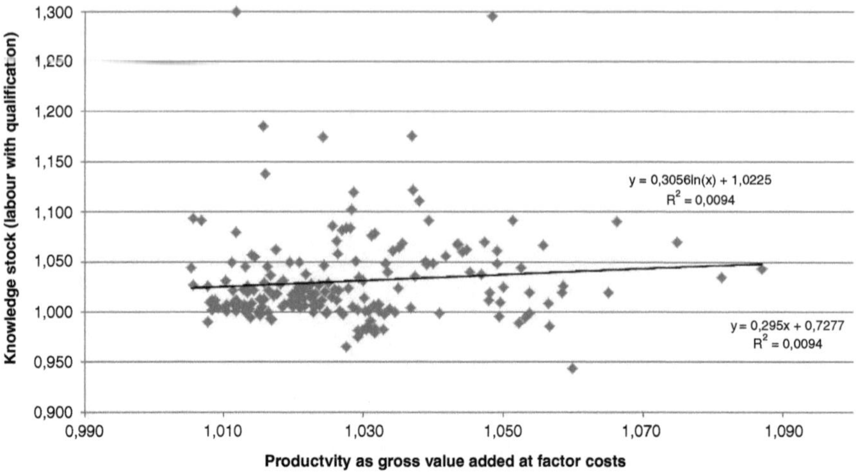

Fig. 6.2 Average dynamics of human capital (labor with tertiary qualifications) and productivity (the rate of regional gross value added (GVA)) at basic prices in the EU regions (1998–2008) (Source: own calculations)

belonging to the same industry can benefit from dynamic externalities and knowledge spillover effects among firms; in other words, it should lead to real learning and adjustment effects. Altogether, Figs. 6.1 and 6.2 show that the process of technological catching up across EU regions is likely to be slow and uncertain, and requires substantial learning and knowledge spillover efforts.

Furthermore, investments in human capital are considered to facilitate knowledge spillovers and the diffusion of new technologies, which increases the productivity of both physical capital and human capital in general. However, as many studies have shown, the relationship between human capital and productivity is not unambiguous (Fuente and Domenech 2006; Vandenbussche et al. 2006; Aghion and Howitt 1992). Indeed, Fig. 6.2 shows that a labor force with tertiary education contributes to increased productivity in the EU Member States NUTS2 regions. However, the correlation trend between the average dynamics of the two variables is less optimistic than one should expect. This, perhaps, can be easily explained by the logic that education itself does not affect workers' abilities and productivity. They obviously learn certain valuable skills at school (engineering, computer science, signalling models); however, their productivity will depend on the use of these skills and the ability to learn from experience.

Nevertheless, this trend in Fig. 6.2 opens a discussion concerning which skills are needed and whether local universities and educational institutions can provide them.

The last two charts in this section discuss the relationship between the growth of GDP per capita, productivity, and employment in knowledge-intensive sectors. Figure 6.3 shows that productivity is one of the key factors that determine economic growth and prosperity in the EU regions. If productivity reflects all the technologies

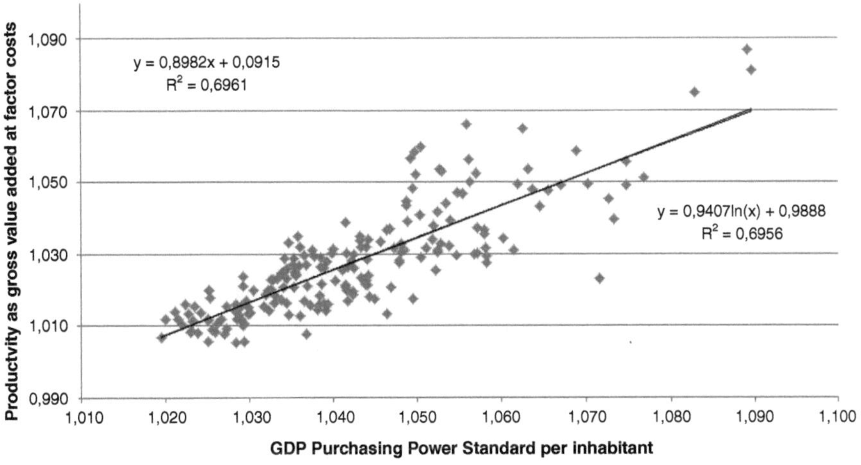

Fig. 6.3 Average dynamics of GDP per capita and rate of productivity (regional gross value added (GVA)) at basic prices in the EU regions (1998–2008) (Source: own calculations)

available at the time and is driven by innovations, then changes in productivity can proxy technological change. One could conclude that technological change leads to an increase in wages and per capita growth of GDP. It also enables the government to collect more income tax revenue and, therefore, it maintains the growth in living standards over the long term.

Furthermore, acceleration of technological advancement and an increase in global competition has generated the process of inter-sectoral reallocation in many EU Member States regions, from low- to high-productivity sectors. Firms reduce their activities in sectors with low technological sophistication and move their resources to sectors with higher technology levels, so overall productivity and income increase. Accordingly, there should be a positive relationship between the share of knowledge-intensive sectors and the dynamics of income. Figure 6.4 displays the log-correlation between average share of high-tech and knowledge-intensive services and GDP per capita measured by PPP in the EU Member States regions in 1998–2008. There is a positive but still low correlation between the variables, which can also be explained by the fact that regions with similar levels of GDP per capita may have distinct knowledge bases and specialize in different high-tech industries. Since productivity gains are not equally distributed across industry level, this relationship should bring different results across the chart for each industry considered. For example, there is rich literary evidence that occupations related to the ICT industry and services generated the highest growth rate of GDP per capita.

To conclude, despite the fact that the EU has a long tradition of education and new knowledge generation, it shows a very modest ability to make EU countries more productive and grow them. The important role of productivity and knowledge-based sectors in improving EU regional prosperity suggests we should

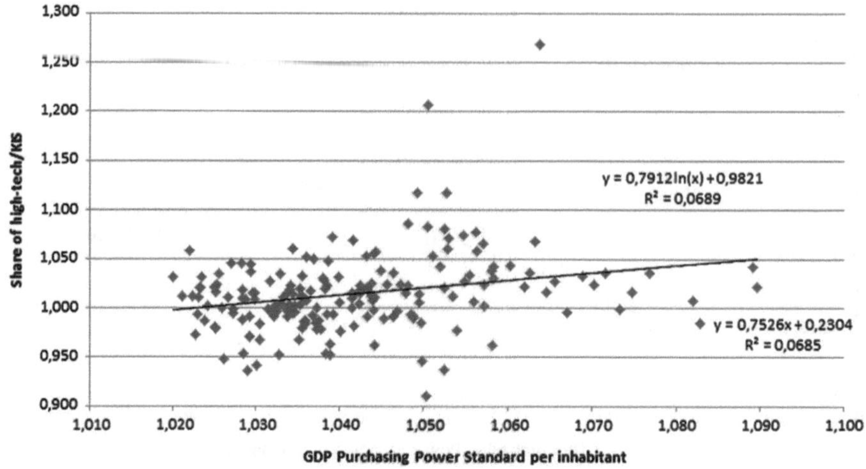

Fig. 6.4 Average dynamics of GDP per capita and share of high-tech and knowledge- intensive services for the EU regions (1998–2008) (Source: own calculations)

carefully examine which knowledge activities drive productivity and the catching-up process of the EU regions, as well as if there are any spatial externalities between them.

6.5 Technological Change and Catching Up in EU Regions. Applied Empirical Analysis

6.5.1 Model Framework

The aim of this section is to check whether the key determinants of endogenous growth theory and new geography can explain the growth patterns of EU Member States regions. More specifically, it aims to answer the following questions: What is the role of the technology gap and technological knowledge spillovers in EU regions' growth and prosperity? What are the key factors explaining the technological gap between the EU regions? What is the role of absorptive capacity and innovative capability of the EU regions in explaining productivity growth? And, finally, what is the role of innovations and high-tech knowledge-intensive services in creating high value-added manufacturing jobs?

In the light of the new growth theories and "catching-up" models and considering existing evidence, the author assumes the technological difference between economies and their capacity to imitate are the main mechanisms of catching up. Technological change represents the factor that shifts the production function. Furthermore, the theoretical rationale draws from the endogenous growth models and localized knowledge spillovers of NEG applied to the EU NUTS2 regional level. The model allows for conditional convergence to a long-run level of productivity relative to the frontier. Similar to the above explained approaches of Fingleton and McCombie (1998) and Pons-Novell and Viladecans-Marsal (1999), the author takes up knowledge-intensive manufacturing in the high-tech sector as the engine of economic growth to find evidence of substantial localized increasing returns to scale in EU manufacturing.

In the spirit of Abramovitz (1986, p. 390), the model will test the role of educational attainment, advanced research, and particularly training skills, in understanding, assimilating, and developing new technology. Like Fingleton (2000) and Greunz (2003), the author assumes that the technological difference between economies and their capacity to imitate are mechanisms of catching up. Moreover, relying on the evidence presented by Coe and Helpman (1995), Guellec and van Pottelsberghe de la Potterie (2004), Griffith et al. (2004) and Cameron et al. (2005) that reveal the important relationship between R&D expenditures, productivity, and income growth on the country level, the author believes that by narrowing innovation (proxied as the R&D gap) the regions will narrow their technological gap. Technological change represents the factor that shifts the production function.

Furthermore, R&D expenditures are assumed to be an important indicator of how technological improvements[2] lead to productivity improvements. This indicator, coupled with the share of the labor force undergoing a lifelong learning process, will proxy the role of the innovative potential of EU regions in productivity growth. The empirical literature has largely ignored the role of lifelong learning in the productivity growth of the EU regions. Lifelong learning is an especially important aspect of productivity growth, particularly in view of the evidence of several studies that providing lifelong learning opportunities encourages the development of innovative skills (Abernathy and Utterback 1978; Gatignon et al. 2002).

In addition, on the basis that many empirical studies that have found a positive relationship between innovations and employment at firm level (Van Reenen 1997; Greenan and Guellec 2001; Harrison et al. 2008), the author attempts to test whether the number of high-tech patent applications influences changes in high-tech employment. Last, but not least, the author aims to investigate the importance of high-tech knowledge-intensive services as a knowledge diffusion channel among companies, and a proxy for the potential effect of outsourcing; the author assumes that the growth of these services will create, capture, and accumulate knowledge in a region and, therefore, will reinforce the cumulative dynamics of innovation. This is especially important, since there is hardly any empirical evidence of the role of high-tech knowledge-intensive services in knowledge-intensive manufacturing employment.

[2] In the author's opinion, based on interviews with experts in several high-tech fields, technological improvements are better proxied with R&D expenditure because in such high-tech fields as aerospace or the ICT, industry many innovations are not being patented.

Finally, considering the fact that technology spillovers from capital investments may be of great importance for the recent EU Member States regions, the present econometric model considers it important to test the influence of capital embodied technology transfer and manufacturing investment on the productivity growth of EU regions.[3]

Following the above argumentation, the author considers the structural system of Eqs. 6.1, 6.2, 6.3, and 6.4.

Equation 6.1 is basically an attempt to answer how much the regional catching-up process depends on the technology gap variable (TG), growth of productivity of its own region (ph), productivity growth of neighboring regions (pg), concentration of knowledge-intensive sectors such as high-tech (S), and on high-tech knowledge-intensive services (hkis). The Eq. 6.1 of the model presupposes that the primary factor that drives GDP per capita growth is productivity per employee (ph). Employees can work less but produce more goods and services and, therefore, they may generate higher incomes and a higher standard of living for everyone. In the case of the variables (S) and high-tech knowledge-intensive services (hkis), one can assume that knowledge-driven growth requires not only a significant rise in the share of knowledge-intensive industries in total manufacturing employment, but also a positive and significant correlation between shares of high tech in the growth rates of GDP.

Equation 6.2 captures the idea that technological differences between regions allow the lagging ones to catch up with the leading ones by imitating their more productive technologies. The size of the income gap depends on the size of the technological gap (T*G*). The level of income will be higher in the technology-leading region. Increased innovation in the leading region will increase the income gap, just as the intensified speed of imitation/technological catching up by the lagging region will contribute to its reduction. The TG is measured at the regional level of real GDP per capita relative to the average of the three best performing EU regions (namely). Furthermore, TG is assumed to be determined by the R&D gap (rdgap), the region's knowledge absorption capacity proxied as human qualifications at the ISCED 5–7 level (HQ), and human resources in science and technology (HRST). Moreover, since technological change represents the factor that shifts the production function, the share of manufacturing employment in high-tech manufacturing is an important input (S).

Furthermore, Eq. 6.3 states that the regional growth rate of productivity (*p*) depends on the growth rate of real GDP per capita (y). This is based on the Kaldor-Verdoorn relationship, according to which productivity growth determines and is determined by the growth of GDP per capita. Productivity growth (p) is also assumed to be positively influenced by manufacturing investment (I) per employee, a main mechanism of the transfer of embodied technologies to a region, the region's

[3] Foreign direct capital flows improve capital and technology endowments of the recipient countries and additionally create technology spillovers to local firms (Saggi 2002; Weresa 2002).

endowment of human qualifications (HQ), regions' innovation propensities (IP), and participation in the lifelong learning process (LL). The existing stock of knowledge and the lifelong learning of regional agents generate higher returns. This happens because innovation is a result of continuous improvements yielded by learning. The more education among the labor force has been accumulated, the higher the knowledge stock is and the better it will be in creating new technologies.

Equation 6.4 assumes that the structure of the productive system (*S*) proxied by employment in high-tech sectors is a positive function of the patenting activity of a region (*P*) and high-tech knowledge-intensive services (HKIS). The latter can play a strategic role in knowledge diffusion, the potential effect of outsourcing, and thus the promotion of the development and growth of high-tech manufacturing industries.

The structural system of equations may be formulated in the following way:

$$y = a_1 + a_2\,TG98 + a_3\,p_h + a_4\,p_g + a_5\,S + a_6\,HKIS\varepsilon \tag{6.1}$$

$$TG = b_1 + b_2S + b_3HQ + b_4HRST + b_5RDGAP + \varepsilon \tag{6.2}$$

$$p = c_1 + c_2I + c_3HQ + c_4IP^4 + c_5LL + \varepsilon \tag{6.3}$$

$$S = d_1 + d_2P + d_3HKIS + \varepsilon \tag{6.4}$$

All the variables are expressed in terms of averages over the period 1998–2008, where

y the growth rate of real GDP per capita expressed in PPS, deflated by the GDP deflator with respect to the price level of 2000.

TG[5] the technology gap, defined in 1998 as $\left(1-\frac{Y}{Y*}\right)$ where $\frac{Y}{Y*}$; it is the regional level of real GDP per capita expressed in PPS, deflated by the GDP deflator with respect to the price level of 2000 relative to the average of the three best performing regions, namely Brussels, Luxembourg and London.

p_h the growth rate of labor productivity in the home region, defined as the ratio of real GDP per employee expressed in PPS, deflated by the GDP deflator with respect to the price level of 2000.

p_g the growth rate of labor productivity of geographical neighbors, defined as the ratio of real GDP per employee expressed in PPS, deflated by the GDP deflator with respect to the price level of 2000.

S the percentage of employment in high-tech manufacturing in total manufacturing employment (DG24.4, DL30, DL32, DL33, DM35.3).

[4] R&D expenditures constitute a better indicator of how increased R&D activity leads to productivity improvements, since in many cases technological improvements are not being patented (e.g. aerospace sector). http://economics.huji.ac.il/facultye/zeira/Innovations8.pdf

[5] TG is introduced in the model in order to present the region's stock of existing knowledge and on its distance to the technological frontier (Fagerberg 1988).

HKIS high-tech knowledge-intensive services, defined as a percent of total services employment in the economy.
HRST human resources in science and technology, defined as a share of the economically active population (15–74 years old).
I manufacturing gross investment in tangible goods per manufacturing employee, expressed in PPS, deflated by the GDP deflator with respect to the price level of 2000.
P_h the number of patent applications to the EPO per 1,000 inhabitants. Patents are attributed to the living place of the inventor.
RDGAP the R&D gap, defined as $(R\&D_h\text{-}R\&D_{best})/R\&D_{best}$, where $R\&D_h$ is the total intramural R&D expenditure per capita at the home region's level by PPP (1998–2008), and $R\&D_{best}$ is the average intramural R&D per capita expenditure of the three best performing regions – Stockholm, Västsverige, and Wien by PPP (1998–2008).
HQ the proportion of highly qualified working age population (25–59 years old) relative to the total working age population (ISCED 5–7).
IP innovation propensity measured as total R&D expenditure, presented as percentage of GDP.
LL the proportion of working age population (25–59 years old) involved in the lifelong learning process as percentage of the total working age population.
εj j = 1...5 a random error term.

The model was tested for the period 1998–2008 on the sample of 186 European regions (NUTS2/3 levels), with a few exceptions: Ireland, Denmark, and Germany, for which data were available only on the national and NUTS1 level. The system of simultaneous equations has been estimated by three-stage least squares (3SLS), without and with regional-specific fixed effects.

6.5.2 Estimates Results

Table 6.1 summarizes the main empirical findings on the estimates. In general, it indicates that the growth rate of GDP per capita is assumed to be a positive function of the initial technological gap. It also shows that poor EU regions should grow more rapidly than the wealthiest ones, and broadly confirms the empirical evidence of Alexiadis and Tomkins (2008) and Alexiadis et al. (2010). More specifically, a 1 % increase of TG generates a 0.11 % growth of real GDP per capita (with a high significance level of 99 %).

Furthermore, the results confirm the expected relationship between real GDP per capita and productivity growth *lpv*. All variables have signs that are consistent with the theory predictions of Kaldor-Verdoorn that output growth would respond positively to productivity growth. In sum, a 1 % increase in productivity level generates a 0.02 % increase in GDP per capita growth. The results show that local level of productivity is statistically significantly related to the region's GDP per

Table 6.1 The role of knowledge flows and technological change in catching up of EU regions (estimated via 3SLS)

	reg1		model3_reg3_we	
Variables	Coeff.	T-stat.	Coeff.	T-stat.
GDP growth				
TG	0.109	(4.26)***	−0.061	(−0.12)
p_h	0.017	(6.57)***	0.017	(3.87)***
p_g(spillover1)	0.002	(0.96)	−0.000	(−0.10)
p_g(spillover2)	−0.000	(−0.13)	0.002	(0.75)
p_g(spillover3)	0.000	(0.34)	0.001	(0.57)
S	−2.100	(−2.62)***	−3.832	(−1.14)
HKIS	0.020	(3.56)***	0.027	(1.07)
_cons	−0.033	(−1.28)	0.006	(1.61)
TG				
S	−0.753	(−0.74)	−0.302	(−1.01)
HQ	0.004	(4.39)***	−0.000	(−0.51)
HRST	−0.01	(−3.23)***	0.002	(4.60)***
RDGAP	0.746	(12.00)***	−0.025	(−0.74)
_cons	0.478	(11.10)***	0.002	(2.47)**
p				
I	15.351	(2.42)**	−49.951	(−1.47)
HQ	0.041	(2.36)**	0.155	(1.92)*
IP	−13.022	(−1.06)	−22.502	(−1.61)
LL	0.054	(2.75)***	0.111	(2.98)***
_cons	0.813	(2.30)**	−0.461	(−5.05)***
S				
P_lag2	−0.000	(−0.09)	0.000	(0.21)
P_lag3	0.000	(1.34)	−0.000	(−0.93)
HKIS	0.008	(23.72)***	0.007	(12.78)***
_cons	0.013	(13.39)***	0.000	(0.27)
N	315.000		315.000	
r2	0.318		−0.820	

Source: own estimations
"reg 1" refers to results without country fixed effects, whereas "model3_reg3_we" considers the country fixed effects). Standard errors in parentheses: ***$p < 0.01$, **$p < 0.05$, *$p < 0.1$

capita growth. The results are consistent with recent studies run by Enflo and Hjertstrand (2009), which focus on labor productivity as a driver of regional growth.

The results for the importance of geographical proximity and influence of technological (productivity) spillovers from neighboring regions on home regions' GDP per capita growth turned out to be insignificant. The outcome suggests that, in order to increase the level of prosperity in lagging regions, regions should increase their home productivity. It also suggests that productivity spillovers may follow strictly industry-specific investment-trade linkages of a particular region, instead of spatially determined knowledge flows via labor mobility (Aitken and Harrison 1999).

The estimates reported in Table 6.1 also indicate that an increase in employment in high-tech industries (S) has a negative and significant influence on GDP per capita growth (an increase of S by 1 % would generate −2.1 % in GDP growth per capita). Although the results turn out to be insignificant when the country fixed effects are considered, this outcome should be interpreted with caution, as there may be several possible explanations for it. One of them is that the EU regions with similar levels of GDP per capita can have distinct knowledge bases and specialize in different high-tech industries which are subject to different business dynamics, and that the impact of these industries on GDP per capita may be controlled for business cycles (see next section).

The findings show that high-tech knowledge-intensive services (HKIS) have a significant but small (0.02) influence on GDP per capita growth, which is somewhat surprising. One way to explain this is the fact that high-tech manufacturing firms increasingly provide services together with their products. Consequently, the input of HKIS to output growth may be greater.

Furthermore, Baily and Gordon (1988) claimed that many high-tech-intensive services, such as computer services, are sold as intermediate goods, so their direct impact on GDP may be small. Therefore, the measurement of the influence of knowledge-intensive services could be an important area for future research.

Taking into consideration the important role that technological infrastructure, especially investments in R&D, plays at the level of knowledge spillovers and technological imitation, it is not surprising that the closing of the technological gap (TG) in Eq. 6.2 is significantly, positively, and highly dependent on narrowing of the R&D gap (*RDGAP*). Indeed, a 1 % drop in *RDGAP* causes a decrease in the *TG* of 0.75 %. This confirms that investment in R&D is a central component of technological catching up and capability to innovate. It also shows that so far, the technology gaps between the EU regions can be exploited through imitation, and also created through innovation.

As previously mentioned in the literature, a region's ability to imitate and produce new knowledge (innovative capability) depends on its knowledge-absorptive capacities, which are proxied by HQ (a proportion of the working-age population with tertiary education) and a proportion of human resources in science and technology degrees (HRST). These factors are assumed to shape the productivity of RIS. Therefore, it is not surprising that the model outcome indicates that HRST is negatively and significantly related to the coefficient of technology-gap (TG). In fact, a 1 % increase in HRST contributes to a 0.01 decrease in the technological gap.

However, the estimates outcome shows that the coefficient of HQ significantly and positively influences technological gap. This outcome is contradictory to common logic that higher education improves a region's capability to absorb new technologies and, therefore, accelerates the technological catching-up process. This finding may be interpreted in the following way: tertiary education, which may cover social sciences, humanities, and technical studies, contributes to the rapid development of human capital in general. However, the emphasis on technical subjects seems to be the most effective way of laying the foundations for technological catching up. While tertiary education is essential, there is also the danger that some regions may ignore the importance of higher education, especially when it comes to mastering science-intensive technologies. The overall prospects for catching up may depend largely on the way in which countries balance their higher education goals. Thus, differently from the standard formulation, where absorptive capacity depends only on human capital, the estimated findings point to the importance of education in S&T in the catching-up process.

As expected from Eq. 6.3, manufacturing investments (I) per employee significantly contribute to foster productivity growth in the EU regions. This seems to be in line with a large body of other studies that show the significant role of investment in economic growth and productivity processes (Cummins et al. 1994; Hassett and Hubbard 1996). It also confirms the relevance of capital investments as a mechanism of the transfer of embodied technologies to regions for the new growth theory. However, since the coefficient (I) covers broad investment components, any attempts to interpret the productivity impact of any type of investment should consider more detailed specifications.

Furthermore, the increase in the proportion of the highly qualified working age population (HQ) and the proportion of the labor force undergoing a lifelong-learning (LL) process turned out to have a significant and positive, but small impact on regional productivity growth (p). Indeed, a 1 % increase in the proportion of the highly qualified working age population will increase productivity by 0.04 % and 0.05 %, respectively. This result is in line with studies by Lucas (1988), Benhabib and Spiegel (1994), and Eaton and Kortum (1996) and confirms the importance of human skills and professionalism for improving the productivity of each worker in the EU regions.

From the last Eq. 6.4 one can observe that the share of an average EU region's employment in the knowledge-intensive manufacturing sector (S) is positively and significantly dependant on the presence of HKIS services. Even though the impact is very small (0.007 and 0.008 when fixed effects are considered), this suggests R&D and ICT services trigger the increase in demand for a variety of high-tech manufactured goods such as computers, cell phones, digital scanners, and optical linkages. Therefore, the close connection between the service and manufacturing sectors is likely to have spillover effects in each of these sectors. The scale and scope of these spillovers depend on each industry sector. For instance, in a study by Francois and Woerz (2007), the authors give an example of the pharmaceutical industry. The increases in the industry's output trigger additional services such as management, advertising, and legal services, and cause additional output from other related industries, such as the chemical industry. In general, the role of services as inputs to other industries was observed by Glasmeier and Howland (1994), Hansen (1993), and Porter (1990, 1998). However, the complexity of technological advancement and the interdisciplinary character of many innovations show there is still much to learn about the mechanisms and sources of knowledge spillovers, which leaves room for further research. The problem of knowledge spillover sources and the role of external knowledge for innovations and technological advancement have been mostly ignored in the economic literature and policy debates to date. Enhancing inter-sectoral knowledge transfer and learning other disciplines may increase the efficiency of RIS and reduce the costs of innovation.

Summarising, empirical evidence based on the 186 EU regions over the period 1998–2008 yields some interesting results. This is especially in reference to the significant, positive, and highly important role of the R&D gap in closing the technological gap and accelerating GDP per capita growth in the EU regions. Another important conclusion refers to the important role of lifelong learning and capital investments as mechanisms of technology transfer into the EU Member States regions. Last, but not least, an important observation from the above findings is that technological catch up depends on education in technical fields, whereas promoting tertiary education may fail to build the absorption and diffusion capacities necessary for the catching-up process.

6.6 Business Cycles and High-tech Employment in the EU Regions

Although the direct economic weight of high-tech industries is relatively small, its impact can be significant for regional economic growth. High-tech industries have a large indirect economic impact on promoting regional growth by inducing innovations on other non-high-tech industries and sectors in the region. ICTs and their application are the prime support needed for the rapid growth and modernisation of various sectors of the economy, as well as creating entirely new industries, e.g. e-commerce. Moreover, there is evidence of significant potential for catching up and convergence in ICT-using manufacturing industries and services, particularly in the EU-10 regions, which have been undergoing a privatization and restructuring process in their manufacturing and service sectors (van Ark and Piatkowski 2004; Kolodko and Piatkowski 2002).

Furthermore, the effects of any new technology introduced on a certain scale extend beyond factories and research centers, influencing the broader scale of social and economic aspects by increasing the standard of living (Eurobarometer 2005). For example, the aerospace technology's greatest contribution – apart from flying – is in satellite communication that delivers information on environmental and security issues; laser technology enabled the shift in storage capacity from vinyl to CDs and DVDs, whereas potential benefits of biotechnology include solving food shortages, improvements in medicine, agriculture, environmental management, etc.[6] These advantages may surpass the rather low impact of biotechnology on labor productivity and cost reductions (Freeman 2003; Nusser et al. 2007; Hopkins et al. 2007). Another issue that arises from technological advancement is the safety

[6] Biotechnology Industry Organization (BIO). "Biotechnology in Perspective," Washington, D.C.: Biotechnology Industry Organization, 1990.

and ethics of the research and its applications, e.g. risk of a genetic engineering accidents in biotechnology or privacy invasion related to the ICT industry. These issues are important as they extend the scope of this book. What is significant from the perspective and purpose of this book is that regions' specialization and participation in global high-tech value chains increase their vulnerability to economic fluctuations (demand shifts and shocks). This is especially true in the case of regions where high-tech industries account for a large share of economic output. Their economies could be more vulnerable to a high-tech contraction. Moreover, since high-tech industries trigger growth in demand for a variety of other high-tech industries and service sectors, as was discussed in the previous section, a synchronous shock in one high-tech sector spreads across a number of related technology industries, e.g. in the computer and semiconductor industries (DeVol 1999). For example, a general downturn in the computer industry in the mid 1990s led to a decline in the demand for semiconductor technology products; as a result, producers responded with lower prices and job cuts (Hart 1986).

The volatility of high-tech industries may be the result of a shortening of life cycles in high-technology products, which can cause major swings in the introduction of new technology products. After a period of rapid growth attributable to new product innovation, the absence of another major innovation may cause a dramatic slowdown, as witnessed in 1985 and 1989 in the computer and semiconductor industries e.g. Route 128 (DeVol 1999).

Altogether, high-tech industries have a strong multiplier effect on the overall economy, and a slowdown can significantly impact the non-high-tech sector, both nationally and regionally. If synchronous shock, especially combined with some other inauspicious development, spreads across a number of related technology industries, an economy-wide recession may occur.

One approach to analyzing the behavior of high-tech industries over the business cycle is to compare the relative dynamics of the high-tech industry and knowledge-intensive service sectors to the overall economic GDP dynamics and total industry and service sector employment trends. Figures 6.5 and 6.6 show that for all EU regions high-tech industry employment seems to be inherently volatile. The changes in high-tech industry employment are even more pronounced for the EU-10 Member States regions, which may introduce risks for the regions that have developed high-tech clusters.

Until the year 2000, the growth of the high-tech market was relatively fast; however, in 2001 the growth rate declined as a result of the global ICT downturn, but in 2002 it rose again. Figure 6.6 shows there is a delayed effect of high-tech downturn on the EU-10 regions' employment. The highest employment drop in high-tech manufacturing was observed in 2002, when most of the EU-15 had already entered the recovery stage.

The recent global financial crisis brought an economic cooling to markets throughout the world. As a result, a demand for total manufacturing and high-tech products and services in the EU area decreased. Due to large cuts in investment expenditures, both by companies and households, these industries, overall, have been deeply affected by this crisis, which resulted in large overcapacity

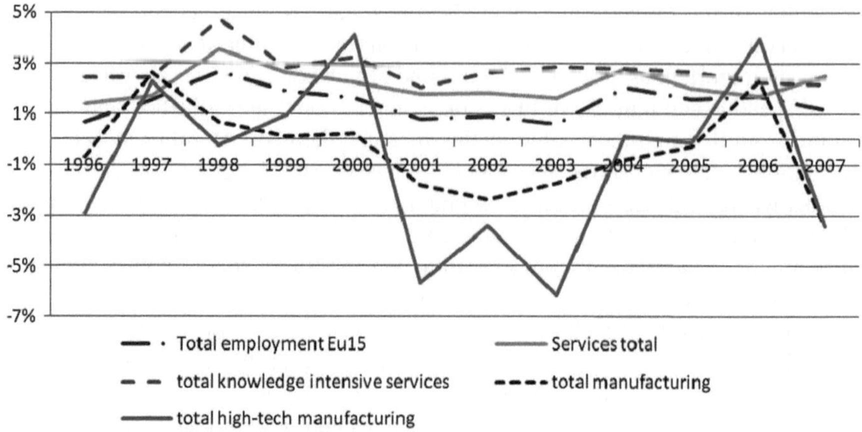

Fig. 6.5 The high-tech sector aligns with the business cycle 1996–2008 for the EU-15 regions (year-by-year dynamics of average values) (Source: own calculations, based on Eurostat data)

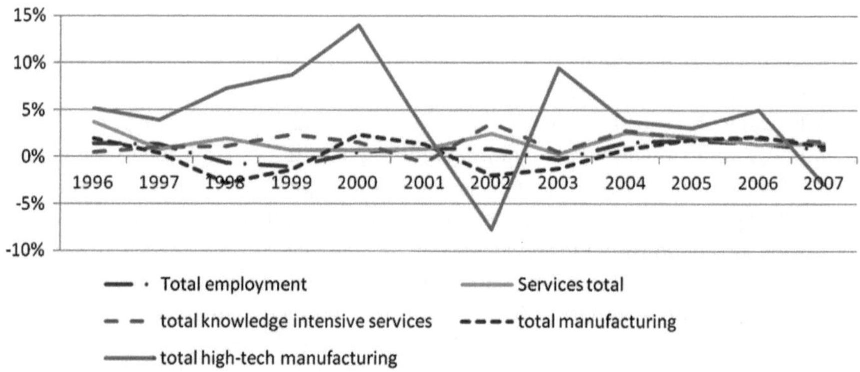

Fig. 6.6 The high-tech sector aligns with the business cycle 1996–2008 for the EU-10 regions (year-by-year dynamics of average values) (Source: own calculations, based on Eurostat data)

(for instance in electronic components) and a decline in profits. It is interesting that the scale of the impact of recent economic crisis on ICT manufacturing employment has been less profound than the downturn in 2002–2003. The study on the longer-term trends in this sector suggests that the ICT sector has become less employment-intensive, which means there is greater pressure on skills requirements to cover broader ground and have greater flexibility (Runiewicz-Wardyn 2006). Similar pressures have been observed in other high-tech industries (e.g. semiconductors and electronics), for which e.g. ICT skills were needed and which did not concern ICT professionals alone (OECD 2009; Dedrick and Kraemer 2008). These observations reveal another challenge to high-tech regions: investment in human capital and continuous upgrading of skills.

The vulnerability of EU high-tech manufacturing employment to the business cycle may have several causes; firstly, the participation of many EU-10 smaller economies in offshore outsourcing[7] activities as part of a bigger value chain. Many companies, especially from new Member States, including Slovenia, Slovakia, the Czech Republic, and Estonia, have become subcontractors for Western European and global manufacturing companies, thanks to their friendly business environment and cheaper knowledge base.

The strong presence of in an important suppliers' network can make the region more attractive for other global players to follow. As a result, the region will further develop its technological infrastructure and improve the cluster environment. However, subcontracting may also pose a risk, especially for the subcontractor. Large firms use subcontractors as "buffers" to stabilize their manufacturing capacity, and during a recession they may reduce the amount of work they contract out. For example, the sharp growth of high-tech exports in Estonia in 2000 resulted from the almost two-fold growth of subcontracting that was driven mainly by contracts allocated by Nordic ICT companies. An overwhelming majority of subcontracting agreements within the Estonian electronics industry was signed for Sweden (55 % in 1995) and Finland (34 %).

However, worldwide difficulties in the ICT sector reduced the volume of subcontracts and the contribution of high-tech exports in 2001, which resulted in a fall in exports of high- technology products by 32 % in 2001; the decline stopped in 2003, when it reached a lower level than in 1999. Although the share of high-tech exports remained quite high throughout this downturn period, it indicated that Estonia's high-tech export was vulnerable (Rajasalu and Laur 2003).

The regions' vulnerability to high-tech cycles can be explained also by the economic dependence on certain suppliers coming from the same region. Therefore, this can explain Estonia's vulnerability to the Finnish and Swedish economic recessions in the early 1990s. It is also worth mentioning the concern of regions dependent on the few international contractors of being dumped by the re-location of high-tech manufacturing to areas of cheaper labor cost, e.g. China, Russia, or India (Tiits et al. 2003).

In sum, high-tech manufacturing may be an important driver of regional growth and employment; on the other hand, high-technology sectors and globally integrated firms expose production processes throughout the economy to both internal and external shocks. When considering the vulnerability of the EU high-tech manufacturing sector to business cycles, policy efforts should focus on reducing regions' dominance on a few large firms or a few industries and promoting a diversified industrial manufacturing value chain.

[7] Subcontracting abroad or offshore outsourcing occur when one firm (the manufacturer or contractor) contracts with another firm (the subcontractor) for a given production cycle. *Handbook on Globalization Indicators,* OECD, 2005.

6.7 Summary and Conclusions

Although an increasing number of empirical studies has paid attention to issues of economic convergence in the EU, the impact of technological gap and knowledge spillovers has received limited attention to date. The following section aims to address this issue by developing a model based on the structural system of equations that enable the identification of the role of the technology gap, absorptive capacity, and innovative capability of the EU regions in explaining productivity growth and the catching-up progress. The empirical evidence based on the 186 EU regions of the EU over the period 1998–2008 emphasizes the significance of new (endogenous) catching-up theories in explaining regional growth in the EU. Broadly, the estimation outcomes show that the technological difference between regions is the mechanism for the catching up of the EU regions. The results for the growth of GDP per capita in the EU regions proved to depend on the initial technology gap (between a home region and the average of the three best performing regions in the EU), labor productivity growth, high-tech knowledge-intensive services, and high-tech manufacturing employment. The last-named also confirms that manufacturing is an important engine of economic growth in regions and in line with Kaldor's first law. The significant and positive role of high-tech knowledge-intensive services confirms their important role in reinforcing the cumulative dynamics of innovation processes in the EU regions.

The estimations results indicate the significant, positive, and highly important role of the R&D gap in closing the technological gap and accelerating GDP per capita growth in the EU regions. The latter is in line with other empirical findings by Coe and Helpman (1995), Guellec and van Pottelsberghe de la Potterie (2004), Griffith et al. (2004), and Cameron et al. (2005), and proves that R&D is an important input-based measure of technological change.

Another important conclusion refers to the role of lifelong learning, which turns out to be positive and significant for productivity growth in the EU regions. The results suggest that lifelong learning skills improve understanding, assimilating, and developing new technological knowledge. This is true with Abramovitz's concept of "social capability" and broadly in line with other empirical findings testing the impact of educational attainment on innovative and technological skills based on the EU-15 and US experience (Fingleton 2000; Greunz 2003, 2004; Abernathy and Utterback 1978; Gatignon et al. 2002).

Last, but not least, the positive and significant role of manufacturing investments in the productivity growth of the EU regions proves that capital-embodied technology transfer capital is very important for increasing the stock of regional knowledge and inducing technological change. It may also suggest that many EU Member States regions are still in transition from the capital-driven to the innovation-driven stage of growth (Porter 1990, 1998), for which technological change is exogenously determined.

6.8 Final Conclusions and Implications

6.8.1 Implications for Theory and Future Research

With regard to the "systemic approach" in explaining regional innovation processes, the research findings show its analytical framework is useful in explaining innovation processes in the EU regions. However, in the case of regions that have recently undergone a systemic transformation, a more dynamic or evolutionary approach to understanding their institutional interlinkages and regional innovation performance is needed. Furthermore, the future framework for the "regional innovation system" (RIS) must consider, on the one hand, the global, complex character of technological knowledge (resulting from technology collaboration networks) and the role of local dynamic externalities, on the other. The literature on the regions' innovation systems is still relatively small, and almost no empirical attempts have been made to incorporate the above elements into a dynamic RIS model. The research findings presented provide clear grounds for further attempts to develop the RIS framework and determine discussion on innovation as a technological, institutional, and evolutionary process.

This book also contributes to the ongoing discussion on the role of space and technological proximity in the process of knowledge spillovers and innovation. Despite the growing number of empirical studies, the role of geographical distance in knowledge spillovers and innovation is still rather controversial. The research evidence of geographical knowledge spillovers confirms the dominant literature on geographical clustering and localized knowledge spillovers, and takes it forward with new outcomes that support the idea that high-tech innovations need to incorporate intra- and extra-regional knowledge sources. This contrasts with the results of past studies on a broader group of patents, which found that only the geographical neighborhood (inter-regional knowledge flows) strongly influences the number of patents in the home region. Consequently, it supports the idea that the innovative performance of regions depends on strengthening "global–local" knowledge interlinkages. The intensity of these interlinkages may be determined by specific fields and their technological proximity. The latter should constitute an important context for regional innovation, technological, and growth processes.

Further comparative research on geographically and technologically mediated knowledge spillovers in the context of changing spatial patterns induced by ongoing integration and globalization processes should reveal fresh directions for new theories of economic geography. The complexity of many technological advances and their interdisciplinary character shows that there is still much to learn about the mechanisms and sources of knowledge spillovers, which opens the door to further research debate.

The research findings show that the presence of dynamic externalities (MAR, Jacobs', and Porter's externalities) varies with the industry and its technological maturity. Agglomeration externalities in the sense of Jacobs' are predominant in the case of most high-tech industries (aerospace, communication, biotechnology, and laser instruments), regardless of their technological maturity. This is explained by the fact that technological advances in these industries converge with the expansion of other industries (especially IT) and therefore require close proximity and interaction with other industries and universities (especially in the case of biotechnology).

The results show that technological advance and globalization have a pervasive influence on the spatial distribution of economic activity. On the one hand, the globalization of production networks and opportunities for offshoring and outsourcing seem to be forcing convergence among regional economies and their technological development; on the other hand, dynamic agglomeration externalities force companies to cluster around the same places. Modern agglomeration theories must, therefore, be tested further for their applicability to the study of how high-tech industries make their location decisions and whether sufficient critical mass is achieved to develop clustering. This is because technological advance in these industries involves interaction between many factors within a complex, dynamic environment, so the weight assigned to the array of location factors will vary with the high-tech industry, the specific knowledge base, and the technological path dependencies. Furthermore, the slight evidence of externalities resulting from outsourcing and reallocating production activities (offshoring) emphasizes the importance of good measurement of their scope and the impact on the spatial distribution of economic activity.

The research findings also show that education and R&D institutions are important drivers of growth and confirm that new (endogenous) economic growth theories are useful in explaining EU regions' growth patterns. This is particularly true of the role of government R&D investments, which appear to play a positive role in promoting quality of education and infrastructure (as in endogenous growth theory). Notwithstanding, the findings are also in line with the technological gap theories (Abramowitz 1979, 1986, 1994) and show that technological change is the result of investment in education (labor force skills), capital stock accumulation (learning by doing), and R&D intensities. The region's capacity for catching up technologically and for long-term growth is determined by its capacity, first, to absorb and adopt externally generated knowledge spillovers and, secondly, to use its endogenous knowledge sources for imitation and innovation efforts.

The results relating to the influence of technological (productivity) spillover from neighboring regions on home regions' GDP per capita growth turned out to be insignificant. This is somewhat surprising. Weak technological spillover between geographical neighboring regions may mean that the rate of technological change does not vary with the region or that the absorptive capacity of local firms to benefit from externalities tends to be weaker in the home region. Considering that the EU regions' growth is determined by a continuous process of integration, the role of technology transfer and knowledge spillover should be further monitored and tested against the existing literature. The research findings suggest that if prosperity in regions lagging behind in development is to improve, their home regions' productivity must also increase. This also suggests that productivity spillover may strictly follow the industry-specific investment-trade linkages of a particular region, instead of spatially determined knowledge flows (via labor mobility).

Furthermore, the research findings emphasize the "broad" Kaldorian view (1960) that learning is closely related to firms' investment activities and knowledge accumulation and that productivity growth determines, and is determined by, the growth of per capita GDP.

In terms of future research, the empirical analysis presented here was based on a relatively short period when many of the EU regions were undergoing difficult years of transition to a market economy. Further investigations using various production models over a longer time span are needed to arrive at a better estimate of the nature and impact of technological change.

The sample of regions considered was disaggregated at NUTS2 and NUTS1 level, which are fairly large. A higher degree of spatial disaggregation would probably have afforded deeper insights into the various mechanisms of knowledge flows, innovation processes, technological externalities, and regional growth dynamics. In addition, the quantitative methods should all be backed up with more detailed insights into the above mechanisms at a given place, for example by case studies. The determination of technological change using input- and output-based measurements must be further enriched with more accurate measures that look at both technological complexity and its economic impact.

6.8.2 *Implications for Policy*

Three decades after the accession of the EU's southern members and almost a decade following the accession of the EU-10 States, the EU still suffers from wide disparities in wealth within and between the acceding Member States, with the former growing faster than the latter. Regions with a per capita GDP significantly above the EU average have higher technological and innovative potential. Based on these findings, the primary aim of regional economic policy for the EU regions (EU-10 Member States and the EU's southern periphery, in particular) should be oriented towards the development of technological infrastructure and local social capital. Public policy should, therefore, focus on stepping up business R&D activities, promoting private–public cooperation, strengthening the industrial base, and improving the labor force's educational qualifications, especially in the S&T disciplines.

Since knowledge absorption and transfer capacities are embodied mainly in the government and university R&D in these regions, public R&D policy can stimulate knowledge spillovers and disseminate them within the productive system by promoting business-university collaborative links, for instance via technology transfer agreements, sponsored research, informal exchanges, and social networks. In addition, economic policy to foster regional competitive advantages should consider the regions' R&D policies in connection with both global competition and regional industrial and knowledge bases.

A relatively high concentration of high-tech employment and innovation activities in the EU-15 (Île de France, Etelä-Suomi, Madrid, Vienna, London, Berlin, and Brussels) and the EU-10 (Prague, Bratislavský, and Közép-Magyarország) capital regions as well as in big metropolitan areas (Bavaria (Munich), Nethersachsen (Hamburg), Baden-Württemberg (Stuttgart and Karlsruhe), Zuid-Holland (The Hague and Rotterdam), Flanders (Antwerp), Utrecht, Groningen, Provence-Alpes-Côte d'Azur (Marseille), Midi-Pyrénées (Toulouse), Aquitaine (Bordeaux), Catalonia (Barcelona), etc.) confirms that high-tech industries benefit mostly from highly agglomerated urban areas, which offer higher knowledge absorption and dissemination capacities. However, if the current trend continues, such externalities may be offset by congestion, higher commuting costs, and pollution. Public policy supporting high-tech industries and industrial clustering should, therefore, be incorporated into long-term regional development strategies. Furthermore, the volatility of high-tech industries' sensitivity to business cycles (resulting from their short life cycles and their global value-chain networks) may be a source of major downturns in the region's production output and economic growth (as in Estonia in 2000, when the country's high-tech exports stemmed from reduced subcontracting from Nordic ICT companies). Considering the high concentration of high-tech manufacturing sectors in the above-mentioned regions, public policy efforts should focus on reducing the regions' dependence on a few large firms or a few industries and promoting a differentiated regional industrial structure.

With regard to RIS efficiency, the empirical findings show that, as far as the patenting equation is concerned, the entire sample of EU regions is positively and significantly influenced by the following explanatory variables: lagged university and business R&D spending, the percentage of tertiary education students, high-tech knowledge-intensive services, and the regions' overall economic conditions (measured in per capita GDP). The latter means that historical and general macroeconomic conditions are important in explaining RIS performance. Hence developing sound macroeconomic policy is essential to foster the economic and social transition to an innovation (or knowledge)-based economy.

The estimates indicate that universities' R&D spending has a significant, positive impact on private lagged business R&D efforts and that business R&D spending feeds back into university R&D efforts. This means the co-location of universities and firms in the same region clearly matters in exploiting knowledge spillovers, especially in industries in which industry R&D, university research, and skilled labor are important inputs. However, the findings show no evidence of business R&D influencing university R&D in the EU-10 regions. This shows that these regions have not developed the endogenous power to generate business-university-industry interactions, and their knowledge flows within their RIS are conditional on public R&D support (unlike most of the EU-15 regions that have developed generative R&D relationships and spontaneous convergence of their industrial and academic worlds). The above consideration shows that it is important to step up support for strong linkages between the various sources of knowledge. From the EU-10 regions' perspective, it is essential to increase formal and informal interactions between universities and business agents by establishing innovation centers, business incubators, S&T parks, free economic zones, etc. The importance of these mechanisms is likely to vary with the industry and the S&T field. The implications for policy may be different in each case.

Government R&D also contributes to fostering university R&D, with a stronger effect in the case of the EU-10 regions. Since university R&D units are funded largely by public resources through grants and contracts, government should have stronger policy instruments encouraging university faculty and students in these regions to disseminate knowledge. Universities should be more actively involved in teaching IPRs in seminars and workshops and by issuing educational material.

The research findings also indicate that private business R&D activities are positively driven by high-tech manufacturing employment and high-tech knowledge-intensive services. They suggest R&D activities generate highly skilled jobs and positive technological externalities in the EU regions. The positive influence of high-tech knowledge-intensive services on private business R&D activity suggests that local companies participate in global knowledge networks through outsourcing and offshoring activities. Wider business R&D networks improve the regions' capacities to create, capture, and accumulate knowledge and therefore reinforce the cumulative dynamics of innovation. Globalization exposes a host country's macroeconomic and business conditions to stronger competitive pressures. Hence, regional economic policy in an enlarged Europe should focus on the development of better technological infrastructure and regional network utilities with a view to a more business-friendly environment for investing in knowledge and R&D, especially in regions with low research capacity and low levels of R&D activity internationalization. Policies geared towards stimulating research mobility flows and R&D networking could improve the quality of local innovation systems. As for the role of agglomeration externalities in explaining the technological dynamics in high-tech industries in the EU regions, the research findings suggest that, on average, innovative activity in most EU high-tech industries depends on Jacobs' externalities resulting from production diversity. The latter suggest that EU high-tech industries are more likely to be located in areas with a history of industrial diversity. This means that public intervention in terms of prioritization of innovation activity and high-tech industrial activity should be restricted. On the other hand, industrial and innovation policy should focus on strengthening what is already established and encouraging the development of emerging clusters by adapting policy to the needs of the particular high-tech industry and its technological maturity. Firms operating in mature industries require a more diversified local industrial structure that provides better access to new knowledge and a fertile environment for the emergence of radical innovations. The latter depends on the quality of the local education system and technological infrastructure. Similarly, in their early stages, industries need public policy support to step up investments in R&D activities and create attractive conditions for entrepreneurship, the provision of venture capital, and cutting red tape.

Finally, with regard to the role of knowledge spillovers and technological change in regional growth, the research findings show that the growth of per capita

GDP in the EU regions depends on the initial technology gap (between a home region and the average of the three best-performing EU regions: Brussels, Hamburg, and Île de France), labor productivity growth, and the percentage of employment in high-tech knowledge-intensive services and high-tech manufacturing industries. Regarding the last variable, the vulnerability of high-tech industries to business cycles should focus policy efforts on reducing the regions' dependence on a few large firms or a few industries and promote a more diversified industrial structure.

With reference to technological gap, the research findings show that closing the technological gap depends on the extent to which the R&D gap in the EU regions can be narrowed. Unlike the standard formulation (where absorptive capacity depends on human capital alone), the findings point to the importance of education in the S&T fields. The research findings show that the technological difference between regions is an important driver of the catching-up process.

Regarding the qualified working age population (with a tertiary education background), the percentage of the labor force undergoing a lifelong-learning process and manufacturing capital investments turn out to have a significant, positive impact on regional productivity growth. Broadly, this means technological change in the EU regions depends on social capital and transfer, and the accumulation of embodied technology capital. Finally, the results show that employment in high and medium high-tech industries is positively and significantly dependent on the presence of high-tech knowledge-intensive services. The latter implies that participation in global production networks (via outsourcing and offshoring activities) can be viewed as an opportunity for the EU regions to upgrade their local productive and innovation systems and move towards higher added-value activities.

Based on these findings, the primary aim of regional economic policy in an enlarged EU should be the promotion of tertiary education and lifelong learning, especially when it comes to mastering science-intensive technology education (natural science, mathematics, information technology, and engineering) as well as innovation management and organization training upon which depend the successful participation of regions in global production networks. Overall, prospects for catching up will depend largely on how regions balance higher education and R&D priorities and place emphasis on the above strategic fields. These results may be regarded as supportive of recent EU regional policy based on the Lisbon and Europe 2020 Strategies of Smart Growth. In particular, they emphasize the importance of targeted interventions to regional innovation processes, matching the support of knowledge excellence, R&D, and technological innovation with the local industrial dimension and the evolutionary path dependence of each technological field.

Bibliography

Abernathy WJ, Utterback JM (1978) Patterns of innovation in industry. Technol Rev 80(7):40–47
Abramovitz M (1979) Rapid growth potential and its realisation: the experience of capitalist economics in the postwar period. In: Malinvaud E (ed) Economic growth and resources: Vol. 1. The major issues. Macmillan, London, pp 1–30
Abramovitz M (1986) Catching up, forging ahead, and falling behind. J Econ Hist 46:386–406
Abramovitz M (1994) The origins of the postwar catch-up and convergence boom. In: Fagerberg J, Verspagen B, von Tunzelmann N (eds) The dynamics of technology, trade and growth. Edward Elgar, Aldershot, pp 21–52
Abreu M, De Groot HLF, Florax R (2004) Spatial patterns of technology diffusion: an empirical analysis using TFP. Tinbergen Institute Discussion Paper (Paper No. TI 2004–079/3), Amsterdam
Acs ZJ, Audretsch DB, Feldman MP (1992) Real effects of academic research. Am Econ Rev 82:363–367
Acs ZJ, Audritsch D, Feldman M (1994) R&D spillovers and recipient firm size. Rev Econ Stat J 76:336–340, MIT Press
Acs Z, Anselin L, Varga A (2002) Patents and innovation counts as measures of regional production of new knowledge. Res Policy J 31:1069–1085, Elsevier
Acs ZJ (2002) Innovation and growth in cities. Edward Elgar, Northampton, MA
Acs Z, Varga A (2002) Introduction to the special issue on regional innovation systems. Int Reg Sci Rev 25(1):3–7
Acs Z, Anselin L, Varga A (2002b) Patents and innovation counts as measures of regional production of new knowledge. Res Policy J 31:1069–1085, Elsevier
Adams JD (1990) Fundamental stocks of knowledge and productivity growth. J Polit Econ 98:673–702
Aghion P, Howitt P (1992) A model of growth through creative destruction. Econometr Econom Soc 60(2):323–351
Agrawal A, Henderson R (2002) Putting patents in context: exploring knowledge transfer from MIT. J Manag Sci 48(1):44–60, INFORMS
Aitken BJ, Harrison AE (1999) Do domestic firms benefit from direct foreign investment? Evidence from Venezuela. J Am Econ Rev 89(3):605–618
Albino V, Garavelli AC, Schiuma G (1999) Knowledge transfer and inter-firm relationships in industrial districts: the role of the leader firm. Technovation 19(1):53–63
Alexiadis S, Tomkins J (2008) Assessing regional convergence under conditions of technology creation and adoption: evidence from an Enlarged Europe, Paper presented at the international conference of regional studies association, 27–29 May, Prague
Alexiadis S (2010) Interregional differences in adoptive abilities: an alternative framework. J Reg Sci Inq 1(2):41–52

M. Runiewicz-Wardyn, *Knowledge Flows, Technological Change and Regional Growth in the European Union*, Contributions to Economics,
DOI 10.1007/978-3-319-00342-9,

Alexiadis S, Ladias C, Polo A (2010) Converging and diverging regions in the EU: implications for regional policy. Reg Sci Inq 2(2):55–70
Alexiadis S, Tomkins J (2010) Technology adoption and club convergence. MPRA paper 21137. University Library of Munich, Munich
Alexiadis S, Ladias C, Polo A (2010) Converging and diverging regions in the Eu: implications for regional policy. Reg Sci Inq 2(2):55–70
Allen KR (2003) Bringing new technology to market. Prentice Hall, Upper Saddle River
Almeida P, Kogut B (1999) Localization of knowledge and the mobility of engineers in regional networks. Manag Sci 45(7):905–917
Amable B (1993) National effects of learning, international specialisation and growth paths. In: Foray D, Freeman C (eds) Technology and the wealth of nations. Frances Pinter, London
Anbil R, Gelman E, Patty B, Tanga R (1991) Recent advances in crew-pairing optimization at American airlines. Interfaces 21:62–74, www.columbia.edu/~dano/courses/4600/lectures/19/fleetsched.pdf
Andersen B (1999) The hunt for S-shaped growth paths in technological innovation: a patent study. J Evol Econ 9:487–526
Anselin L (1988) Spatial econometrics: methods and models. Kluwer Academic, Boston
Anselin L, Florax R (1995) Small sample properties of tests for spatial dependence in regressionmodels: some further results. In: Anselin L, Florax R (eds) New directions in spatial econometrics. Springer, New York, pp 75–95
Anselin L, Varga A, Acs Z (1997) Local geographic spillovers between university research and high technology innovations. J Urban Econ 42(3):422–448
Anselin L, Varga A, Acs Z (2000) Geographic spillovers and university research: a spatial econometric perspective. Growth Change 31(4):501–515
Antonelli C, Geuna A, Steinmueller WE (2000) Information and communication technologies and the production, distribution and use of knowledge. Int J Technol Manag 20(1–2):72–94
Antonelli C, Fassio C (2012) Academic knowledge and economic growth: are scientific fields all alike? Università di Torino & BRICK working papers Collegio Carlo Alberto
Archibugi D, Michie J (1997) Technological globalisation or national systems of innovation? Futures 29(2):121–137
Arora A, Fosfuri A, Gambardella A (2001) Markets for technology and their implications for corporate strategy. Ind Corp Change 10(2):419–451
Arora A, Merges R (2004) Specialized supply firms, property rights and firm boundaries. J Ind Corp Change 13(3):451–475
Arrow KJ (1962) Economic welfare and the allocation of resources for invention. In: Nelson RR (ed) The rate and direction of inventive activity. Princeton University Press, New Jersey
Asheim BT, Coenen L (2005) Knowledge bases and regional innovation. Systems: comparing Nordic clusters. Res Policy 34(8):1173–1190
Asheim B, Gertler MS (2005) Regional innovation systems and the geographical foundations of innovation. In: Fagerberg J, Mowery D, Nelson R (eds) The Oxford handbook of innovation. Oxford University Press, Oxford, pp 291–317
Asheim BT et al (2003) Regional innovation policy for small-medium enterprises. Edward Elgar, Cheltenham/Northampton
Asheim B, Coenen L, Moodysson J, Vang J (2005) Regional innovation system policy: a knowledge-based approach. CIRCLE electronic working paper no. 2005/13. CIRCLE, Lund
Audretsch DB (1998) Agglomeration and the location of innovative activity. Centre for Economic Policy Research, London
Audretsch DB, Stephan PE (1996) Company – scientist locational links: the case of biotechnology. Am Econ Rev 86(3):641–652
Audretsch DB et al (1996) Start-up size and industrial dynamics: some evidence from Italian manufacturing. WZB, Berlin
Audretsch DB, Feldman MP (2004) Knowledge spillovers and the geography of innovation. In: Henderson V, Thisse JF (eds) Handbook of regional and urban economics: cities and geography. Elsevier Science Publisher, pp 2713–2739

Audretsch DB, Keilbach M (2007) The theory of knowledge spillover entrepreneurship. J Manag Stud 44(7):1242–1254, Wiley Pub

Audretsch DB, Sanders M (2009) Technological innovation, entrepreneurship and development, UNU-MERIT working paper series 052. United Nations University, Maastricht

Autant-Bernard C (2001) The geography of knowledge spillovers and technological proximity. Econ Innov New Technol J 10(4):237–254

Autant-Bernard C (2006) R&D lab location. Evidence from the French case, CREUSET University of Saint-Etienne. http://hal.archives-ouvertes.fr/docs/00/06/65/23/PDF/

Autant-Bernard C, Billand P, Frachisse D, Massard N (2007a) Social distance versus spatial distance in R&D cooperation: empirical evidence from European collaboration choices in micro and nanotechnologies. Pap Reg Sci 86:495–519, Springer

Autant-Bernard C, Billand P, Frachisse P, Massard N (2007b) Social distance versus spatial distance in R&D cooperation: empirical evidence from European collaboration choices in micro and nanotechnologies. Pap Reg Sci 86:495–519, Springer

Autor DH, Katz LF, Krueger AB (1997) Computing inequality: have computers changed the labor market? Princeton University: Industrial Relations Section Working Paper No 377

Badinger H, Müller W, Tondl G (2002) Regional convergence in the European Union (1985–1999): a spatial dynamic panel analysis. HWWA discussion papers 210, Hamburg

Badinger H, Tondl G (2005) The factors behind European regional growth: trade, human capital and innovation. Jahrbuch für Regionalwissenschaft 25:67–89

Bagchi-Sen S (2007) Strategic considerations for innovation and commercialization in the US biotechnology sector. Eur Plann Stud 15(6):753–766

Baily M, Gordon R (1988) The productivity slowdown, measurement issues, and the explosion of computer power. Brook Pap Econ Act 2:347–431

Bajerski A (2008) Sieciowe powiązania kadrowe ośrodków akademickich w Polsce. Przegląd Geogr 80(2):289–300

Banbury CM, Mitchell W (1995) The effect of introducing important incremental innovations on market share and business survival. Strateg Manag J 16(S1):161–182

Bania N, Calkins L, Dalenberg R (1992) The effects of regional science and technology policy on the geographic distribution of industrial R&D laboratories. J Reg Sci 32:209–228

Bania N, Eberts RW, Fogarty MS (1993) Universities and the startup of new companies: can we generalize from Route 128 and Silicon Valley. Rev Econ Stat 75(4):761–766

Baptista R, Swann GMP (1998) Do firms in clusters innovate more? J Res Policy 27:525–540, Springer

Baptista R (2000) Do innovations diffuse faster within geographical clusters? Int J Ind Organ 18:515–35

Baptista R (2001) Geographical clusters and innovation diffusion. Technol Forecast Soc Change 66(1):31–46

Barney J (1991) Firm resources and sustained competitive advantage. J Manag 17(1):99–120

Bar-Shalom A, Cook-Deegan RM (2002) Patents and innovation in cancer therapeutics: lessons from Cell Pro. Milbank Q 80(4):637–676

Bartel AP, Sicherman N (1999) Technological change and wages: an interindustry analysis. J Polit Econ 107:285–325

Bassinini A, Scarpetta S (2002) Does human capital matter for growth in OECD countries? A pooled mean group approach, Elsevier: J Econ Lett 12

Bathelt H et al (2004) Clusters and knowledge: local buzz, global pipelines and the process of knowledge creation. Prog Hum Geogr 28(1):31–56

Baum JAC, Silverman BS (1998) Alliances, patents and competitive dynamics in Canadian biotechnology. Adm Sci Assoc Canada, Adm Sci Q 19(6):61–70

Beal B, Gimeno J (2002) Geographic agglomeration, knowledge spillovers and competitive evolution. INSEAD Working paper, 2001/26/SM

Beckmann M (2000) Interurban Knowldhe Networks. In: Batten D et al (eds) Learning, innovation and the urban evolution. Kluwer Academic, London

Bellak C, Narula R (2008) EU enlargement and consequences for FDI assisted industrial development. Economics and management discussion papers 2008–69. Henley Business School, Reading University

Belussi F, McDonald F, Borrás S (2002) Industrial districts: state of the art review. Research report, Project West–East id: industrial districts re-location processes; identifying policies in the perspective of the European Union enlargement

Benhabib J, Spiegel M (1994) The role of human capital in economic development: evidence from aggregate cross-country data. J Monetary Econ 34(2):143–173, Elsevier

Benhabib J, Spiegel MM (2005) Human capital and technology diffusion. In: Aghion P, Durlauf SN (eds) Handbook of economic growth. Elsevier, Amsterdam, pp 935–966

Berman E, Bound J, Griliches Z (1994) Changes in the demand for skilled labor within U.S. manufacturing: evidence from the annual survey of manufactures. Q J Econ 109(2):367–397

Bertolotti M (2005) The history of the laser. Institute of Physics Publishing, Bristol

Bessant J, Rush H (1995) Building bridges for innovation: the role of consultants in technology transfer. Res Policy 24(1):97–114

Bessen J, Hunt RM (2003) An empirical look at software patents. Federal Reserve Bank of Philadelphia, Philadelphia

Bessen J, Maskin E (2000) Sequential innovation, patents, and imitation. Massachusetts Institute of Technology, Cambridge, MA

Bettis RA, Hitt M (1995) The new competitive landscape. Strateg Manag J 16:6–19

Bierly P, Chakrabarti A (1996) Technological learning, strategic flexibility, and new product development. IEEE Trans Eng Manag 43(4):368–380

BIO (1990) Biotechnology in perspective. Biotechnology Industry Organization (BIO), Washington, DC

Blanchflower DG, Oswald AJ, Millward N (1991) Unionism and employment behavior. Econ J 102:815–834

Blumenthal D, Campbell EG, Anderson MS, Causino N, Louis KS (1997) Withholding research results in academic life science: evidence from a national survey of faculty. JAMA 277:1224–1228

Blumenthal D, Campbell EG, Gokhale M, Yucel R, Clarridge B, Hilgartner S, Holtzman NA (2006) Data withholding in genetics and other life sciences: prevalences and practices. J Acad Med 81(2):137–145

Bogdanienko J, Kuzel M (2008) Znaczenie wymiany wiedzy w procesie współpracy przedsiębiorstw transnarodowych. Problemy Zarządzania 2(20):51–70

Bogliacino F, Pianta M (2010) Innovation and employment. A reinvestigation using revised pavitt classes. Res Policy 39(6):799–809

Borensztein E, De Gregorio J, Lee J-W (1998) How does FDI affect economic growth. J Int Econ 45(1):115–135, Elsevier

Boschma RA, Frenken K (2006) Why is economic geography not an evolutionary science? Towards an evolutionary economic geography. J Econ Geogr 6:273–302, Oxford Journals

Boschma RA, Martin R (eds) (2010) The handbook of evolutionary economic geography. Edward Elgar, Cheltenham

Bosco MG, Brugnoli A (2010) Regional efficiency, innovation and productivity. Cached, RSA conference proceedings, Pécs

Bottazi L, Peri G (2000) Innovation and spillovers: evidence from European regions, vol 340, CESifo Working Papers. CES/Ifo, Munich, pp 1–52

Bottazzi L, Peri G (1999) Innovation, demand and knowledge spillovers: theory and evidence from European regions, Working Papers 153, IGIER (Innocenzo Gasparini Institute for Economic Research), Bocconi University

Bottazzi L, Peri G (2005) The international dynamics of R&D and innovation in the short and in the long run, vol 11524, NBER Working Papers. National Bureau of Economic Research, Cambridge

Boufaden N, Plunket A (2007) Proximity and innovation: do biotechnology firms located in the Paris region benefit from localized technological externalities? Ann Econ Stat ADRES 87–88:197–220

Bovee CL, Thill JV (1992) Business communication today. McGraw-Hill, New York

Braczyk HJ, Cooke P, Heidenreich M (1998) Regional innovation systems. The role of governance in a globalized world. UCL Press, London

Broekel T, Boschma R (2010) Knowledge networks in the Dutch aviation industry: the proximity paradox. J Econ Geogr, Res Policy 38:746–755

Breschi S, Cusmano L (2004) Unveiling the texture of a European research area: emergence of oligarchic networks under EU framework programmes. Int J Technol Manag 27(8):747–772

Breschi S, Lissoni F (2001) Knowledge spillovers and local innovation systems: a critical survey. Ind Corp Change 10(4):975–1005

Breschi S, Lissoni F (2003) Mobility and social networks: localised knowledge spillovers revisited. KITeS working paper no. 142, KITeS, Centre for Knowledge, Internationalization and Technology Studies, Universita' Bocconi, Milano, Revised Mar 2003

Breschi S, Lissoni F, Montobbio F (2007) The scientific productivity of academic inventors: new evidence from Italian data. Econ Innov New Technol 16(2):101–118

Bresnahan TF, Gambardella A (2004) Building high-tech clusters: silicon valley and beyond. Cambridge University Press, Cambridge/New York

Breznitz SM, Anderson W (2006) Boston metropolitan area biotechnology cluster. Can J Reg Sci 28(2):249–264

Brockhoff K, Gupta A, Rotering C (1991) Inter-firm R&D cooperations in Germany. Technovation J 11(4):219–229, Elsevier

Brockhoff K, Teichert T (1995) Cooperative R&D and partners' measures of success. Int J Technol Manag 10(1):111–123, Special Issue on the Management of Technological Flows Across Industrial Boundaries

Bromberg J (1991) The laser in America, 1950–1970. MIT Press, Cambridge

Brulhart M (2000) Evolving geographical specialisation of European manufacturing industries. Cahiers de Recherches Economiques du Dapartement d'Econometrie et d'Economie politique (DEEP), Université de Lausanne

Bun MJG, El Makhloufi A (2007) Dynamic externalities, local industrial structure and economic development: panel data evidence for Morocco. Reg Stud 41(6):823–837

Bunnell T, Coe N (2001) Spaces and scales of innovation. Prog Hum Geogr J 25:569–589, SAGE Pub

Cairncross F (1997) The death of distance. Harvard Business Press, Cambridge

Caloghirou Y, Hondroyannis G, Vonortas N (2003) The performance of research partnerships. Manage Decis Econ J 24(2–3):85–99, Wiley

Caloghirou Y, Kastelli I, Tsakanikas A (2004) Internal capabilities and external knowledge sources: complements or substitutes for innovative performance? Technovation 24(1):29–39

Cameron G, Proudman J, Redding S (2005) Technological convergence, R&D, trade and productivity growth. Eur Econ Rev 49(3):775–807, Elsevier

Campbell J, Hlavka DL, Welton EJ, Flynn CJ, Turner DD, Spinhirne JD, Scott VS, Hwang IH (2002) Full-time, eye-safe cloud and aerosol lidar observation at atmospheric radiation measurement program sites: instruments and data processing. J Atmos Oceanic Technol 19:431–442

Campos B (2002) Bologna and initial teacher education in Portugal, Teacher Education Policy in Europe (TEPE) network. http://tepe.files.wordpress.com/2012/01/tepe_monograph_2010.pdf

Caniels MJ (1997) The geographic distribution of patents and value added accross european regions, Working paper (Août) 10. UNU-MERIT, Maastricht

Capellen A, Castellacci F, Fagerberg J, Verspagen B (2003) The impact of EU regional support on growth and convergence in the European. Union J Common Mark Stud 41(4):621–644

Capello R (1999) SME clustering and factor productivity: a milieu production function model. Eur Plan Stud 7(6):719–735

Capello R (2001) Urban innovation and collective learning: theory and evidence from five metropolitan cities in Europe. In: Fischer MM, Froehlich J (eds) Knowledge, complexity and innovation systems. Springer, Berlin/Heidelberg/New York

Capello R, Camagni R (2012) Regional innovation patterns and the EU regional policy reform: towards smart innovation policies. Paper presented at the 52th ERSA conference, held in Bratislava, 21–24 Aug 2012
Capello R, Faggian A (2005) Collective learning and relational capital in local innovation processes. Reg Stud 39(1):75–87
Capello R, Olechnicka A, Gorzelak G (2012) Universities, cities and regions. Loci for knowledge and innovation creation. Routledge, New York
Capron H (2002) Transition towards the knowledge-based economy: growth potential and learning regions, vol 2, Belgian report on science, technology and innovation. Belgian Federal Office of Scientific, Technical and Cultural Affairs, Brussels
Capron H, Cincera M, Rojo J (2003) Evaluation methodologies for macro-level analysis. ULB Institutional Repository 2013/881, Brussels
Carlsson B, Jacobsson S, Holmén M, Rickne A (2002) Innovation systems: analytical and methodological issues. Res Policy 31(2):233–245
Castellani D, Pieri F (2011) R&D offshoring and the productivity growth of European regions. Working Papers 1120, Department of Applied Economics II, Universidad de Valencia
Castells M (1996) The rise of the network society. Blackwell, Oxford
Castells M (2001) The internet galaxy: reflections on the internet, business and society. Oxford University Press, Oxford
Castiaux A (2007) Radical innovation in established organizations: being a knowledge predator. J Eng Technol Manag 24(1–2):36–52
Caulfield T et al (2006) Evidence and anecdotes: an analysis of human gene patenting controversies. J Nat Biotechnol, 1091 http://www.wulaw.wustl.edu/faculty/documents/kieff/Articles/CaulfieldCookDeeganKieffWalsch AnalysisofHumanGenePatents.pdf
Chaney PK, Devinney TM, Winer RS (1991) The impact of new product introductions on the market value of firms. J Bus 64(4):573–610
Cheshire P, Magrini S (2000) Endogenous processes in European regional growth: convergence and policy. Growth Change 31(4):455–479, Gatton College of Business and Economics, University of Kentucky
Christensen CM, Bower JL (1996) Customer power, strategic investment, and the failure of leading firms. Strateg Manag J 17(3):197–218
Ciccone A (2002) Agglomeration effects in Europe. Eur Econ Rev 46(2):213–227, Elsevier
Cincera M (2003) Comparison of regional approaches to Foster innovation in the European Union: the case of Flanders. ULB Institutional Repository 2013/925, Brussels
Cincera M, Capron H (2001) The assessment of innovation systems from an institutional perspective. ULB Institutional Repository 2013/891, Brussels
Clarysee B et al (2005) Spinning out new ventures: a typology of incubation strategies from European research institutions. J Bus Ventur 20:183–216
Cockburn I, Henderson R (1998) Absorptive capacity, coauthoring behavior, and the organization of research in drug discovery. J Ind Econ 46(2):157–182
Coe DT, Helpman E (1995) International R&D spillovers. NBER working papers 4444. National Bureau of Economic Research, Cambridge, MA
Cohen WM, Levinthal DA (1989) Innovation and learning: the two faces of R&D. Econ J 99:569–596
Cohen WM, Levinthal DA (1990) Absorptive capacity: a new perspective on learning and innovation. Adm Sci Q J 35:128–152, SAGE Pub
Cohen WM, Nelson RR, Walsh JP (2000) Protecting their intellectual assets: appropriability conditions and why U.S. manufacturing firms patent (or not). National Bureau of Economic Research working paper 7552, Cambridge, MA
Constantelou A, Tsakanikas A, Caloghirou Y (2004) Inter-country technological linkages in European framework programmes: a spur to European integration? Int J Technol Manag 27:773–790
Cooke P, Uranga MJ, Etxebarria G (1997) Regional innovation systems: institutional and organizational dimensions. Res Policy J 26:475–491, Elsevier

Cooke P, Uranga MJ, Etxebarria G (1998) Regional systems of innovation: an evolutionary perspective. Environ Plann J 30:1563–1584, Taylor & Francis
Cooke P, Boekholt P, Tödtling F (2000) The governance of innovation in Europe. Pinter, London
Cooke P (2004) Evolution of regional innovation systems – emergence, theory, challenge for action. In: Cooke P et al (eds) Regional innovation systems, 2nd edn. Routledge, London, pp 1–18
Cooke P (2005) Knowledge, innovation and the competitiveness of cities. In: Buck N et al (eds) Changing cities. Palgrave, London
Cooke P, Morgan K (1993) The network paradigm: new departures in corporate and regional development. Environ Plann D Soc Space 11(5):543–564
Cooke P, Morgan K (1998) The associational economy: firms, regions, and innovation. Oxford University Press, Oxford
Cooke P et al (2002) Developing a regional innovation strategy for Northern Ireland. Northern Ireland Economic Development Office, Belfast
Cooke P, Ehret O (2009) Proximity and procurement: a study of agglomeration in the Welsh aerospace industry. J Eur Plann Stud 17(4):549–567
Cooper RG, Edgett SE (2010) Developing a product innovation and technology strategy for your business. Res Technol Manag 53(3):33–40
Cooper AC, Folta TB, Woo C (1995) Entrepreneurial information search. J Bus Ventur 10 (2):107–120
Cornwall J (1976) Diffusion, convergence and Kaldor's law. Econ J 86:307–314
Cortright J, Mayer H (2001) Tech specialization: A comparison of high technology centers. The Brookings Institution, Center on Urban and Metropolitan Policy, Washington, DC
Cortright J, Mayer H (2002) Signs of life: the growth of biotechnology centers in the United States. The Brookings Institution Center on Urban and Metropolitan Policy, Washington, DC
Costa DL, Kahn ME (2000) Power couples: changes in the locational choice of the college educated, 1940–1990. Q J Econ 115(4):1287–1315, MIT Press
Cowan R, Foray D (1997) The economics of codification and the diffusion of knowledge. MERIT, Maastricht
Cowan R, Jonard N (2003) Heterogenous agents, interactions, and economic performance. Springer, Berlin
Cowan R, David P, Foray D (2000) The explicit economics of knowledge codification and tacitness. Ind Corp Change 9(2):211–253
Crane D (1972) Invisible colleges: the diffusion of knowledge in scientific communities. Chicago University Press, Chicago
Crescenzi R, Rodrıguez-Pose A, Storper M (2007) The territorial dynamics of innovation: a Europe–United States comparative analysis. J Econ Geogr 7:673–709
Cumbers A, MacKinnon D, Chapman K (2003) Innovation, collaboration and learning in regional clusters: a study of SMEs in the Aberdeen oil complex. Environ Plan A 34:1689–1706
Cummins JG, Trevor SH, Kevin AH (1994) Accounting standards, information flow, and firm investment behavior. NBER Working Papers 4685, National Bureau of Economic Research
Czarnitzki D, Spielkamp A (2000) Business services in Germany: bridges for innovation. ZEW discussion paper no. 00–52, Mannheim
Dasgupta P, David P (1994) Toward a new economics of science. J Res Policy 23(5):487–522, Elsevier
David PA (1994) Why are institutions “the carriers” of history? Path-dependence and the evolution of conventions, organizations and institutions. Struct Change Econ Dynam 5(2):205–220
Davis L, Larsen MT, Lotz P (2011) Scientists' perspectives concerning the effects of university patenting on the conduct of academic research in the life sciences. J Technol Transfer 36 (1):14–37
Debackere K, Veugelers R (2005) The role of academic technology transfer organizations in improving industry science links. Res Policy J 34:321–342, Elsevier
DeBresson C, Amesse F (1991) Networks of innovators: a review and introduction to the issue. Res Policy 20:363–379

Decarolis DM, Deeds DL (1999) The impact of stocks and flows of organizational knowledge on firm performance: an empirical investigation of the biotechnology industry. Strateg Manag J 20 (10):953–968

Dedrick J, Kraemer K (2008) Impact of globalization and offshoring on engineering employment in the personal computing industry. In: The offshoring of engineering: facts, unknowns, and potential implications; National Academy of Engineering (NAE). Committee on the Offshoring of Engineering, National Academies Press, Washington DC

Denison E (1979) Accounting for slower economic growth. The Brookings Institution, Washington, DC

Denny M, Fuss M (1983) The effects of factor prices and technological change on the occupational demand for labour: evidence from Canadian telecommunications. J Hum Resour XVII (2):161–176, AFAS

Desrochers P (2002) Industrial ecology and the rediscovery of inter-firm recycling linkages: historical evidence and policy implications. Ind Corp Change 11(5):1031–1057

DeVol R (1999) America's high-technology economy: growth, development, and risks for Metropolitan areas. Milken Institute, Santa Monica

Dewar RD, Dutton JE (1986) The adoption of radical and incremental innovations: an empirical analysis. Manag Sci 32(11):1422–1433

DiGregorio D, Shane S (2003) Why do some universities generate more start-ups than others? Res Policy 32(2):209–227

Dijkstra BR, Manderson E, Lee T-Y (2011) Extending the sectoral coverage of an international emissions trading scheme. J Environ Resour Econ 50:243–266, Springer

Dodgson M (2000) The management of technological innovation: an international and strategic approach. Oxford University Press, Oxford

Doloreux D (2002) Characterizing the regional innovation systems in Sweden: a tentative typology based on a description of responses to the community innovation survey II. Nordisk Samhällsgeografisk Tidskrift Journal 34(1):69–92

Doloreux D, Parto S (2005) Regional innovation system: current discourse and unresolved issues. Technol Soc 27:133–153

Dominicis L de, Florax RJGM, Groot HLF de (2011) Regional clusters of innovative activity in Europe: are social capital and geographical proximity the key determinants? Tinbergen institute discussion paper, Timbergen Instituut, Amsterdam

Döring T, Schnellenbach J (2006) What do we know about geographical knowledge spillovers and regional growth? A survey of the literature. Reg Stud J 40:375–395, Taylor&Francis

Dosi G (1982) Technological paradigms and technological trajectories. Res Policy 11(3):147–162

Draganinska S, Wink R (2005) Interregional and international knowledge flows in medium technology sectors: the role of formal and informal institutions. Paper presented at the 45th ERSA Congress, Amsterdam, August 2005

Drucker PF (1985) Innovation and entrepreneurship: practice and principles. Harper and Row, New York

Dunnewijk T, Hollanders H, Wintjes R (2008) Benchmarking regions in the enlarged Europe: diversity in knowledge potential and policy options. In: Nauwelaers C, Wintjes R (eds) Innovation policy in Europe; measurement and strategy. Edgar Elgar, Cheltenham

Duranton G, Puga D (2001) Nursery cities: urban diversity, process innovation, and the life cycle of products. J Am Econ Rev 91(5):1454–1477, American Economic Association

Easingwood C, Moxey S, Capleton H (2006) Bringing high technology to market: successful strategies employed by the worldwide software industry. J Prod Innov Manag 23(6):498–511

Eaton J, Kortum S (1996) Trade in ideas: patenting and productivity in the OECD. J Int Econ 40:251–278, Elsevier

Edquist C, Hommen L, McKelvey M (2001) Innovation and employment: process versus product innovation. Edward Elgar, Cheltenham

Edwards ME (2005) Regional and urban economics and economic development. Theory and methods. Auerbach, New York

Ejermo O, Gråsjö U (2008) The effects of R&D on regional invention and innovation. CIRCLE Electronic Working Papers 2008/3, Lund University, CIRCLE - Center for Innovation, Research and Competences in the Learning Economy
Encaoua D, Guellec D, Martinez C (2003) The economics of patents: from natural rights to policy instruments. National Bureau of Economic Research, http://www.nber.org/CRIW/papers/encaoua.pd (2003.124)
Enflo K, Hjertstrand P (2009) Relative sources of european regional productivity convergence: a bootstrap frontier approach. J Reg Stud 43(5):643–659, Taylor and Francis
Entorf H, Pohlmeier W (1990) Employment, innovation and export activity. Evidence from firm-level data. Surveys and applications. Blackwell, Oxford, pp 394–415
ESPON (2012) The sixth volume of ESPON Territorial Observations focuses on regions and cities in the global economy. http://www.espon.eu/export/sites/default/Documents/Publications/TerritorialObservations/TO6_May2012/TO-6_Regions_and_cities_in_the_global_economy.pdf. Accessed May 2012
Etzkowitz H, Leydesdorff L (2000) The dynamics of innovation: from national systems and 'mode 2' to a triple helix of university-industry-government relations. Res Policy 29:109–123
Eurobarometer (2005) Eurobarometer surveys. http://ec.europa.eu/public_opinion/archives/eb_arch_en.htm
Eurobarometer (2007) Eurobarometer surveys. http://ec.europa.eu/public_opinion/archives/eb_arch_en.htm
Eurostat (2011) Regional Yearbook 2011. http://bookshop.europa.eu/en/eurostat-regional-yearbook-2011-pbKSHA11001/
Eurostat (2005, 2006, 2008, 2011), General and regional statistics, The Statistical Office of the European Union, http://epp.eurostat.ec.europa.eu/portal/page/portal/statistics/themes
EUROSTAT (2010) Eurostat yearbook 2008: Europe in figures. Office for Official Publications of the Luxembourg
EUROSTAT (2012) Eurostat yearbook 2008: Europe in figures. Office for Official Publications of the European Communities, Luxembourg
Evangelista R, Iammarino S, Mastrostefano V, Silvani A (2002) Looking for regional systems of innovation: evidence from the Italian innovation survey. Reg Stud J 36(2):173–186
Evangelista R, Savona M (2003) Innovation, employment and skills in services. Firm and sectoral evidence. Struct Change Econ Dyn 14(4):449–474
Evangelista R, Vezzani A (2010) The economic impact of technological and organizational innovations. A firm-level analysis. Res Policy J 39:1253–1263, Elsevier
Fabrizio KR, Di Minin A (2008) Commercializing the laboratory: faculty patenting and the open science environment. J Res Policy 37(5):914–931, Elsevier
Fagerberg J, Hansson P, Lundberg L, Melchior A (eds) (1997) Technology and international trade. Edward Elgar, Cheltenham
Fagerberg J et al (2005) The Oxford handbook of innovation. Oxford University Press, Oxford
Feldman MP (1994) The geography of innovation. Kluwer Academic, Boston
Feldman MP (2001) The enterpreneurial event revisited: firm formation in a regional context. Ind Corp Change 10(4):861–891
Feldman MP, Audresch BD (1996) Innovative clusters and the industry life cycle. Rev Ind Organ 11(2):253–273
Feldman MP, Florida R (1994) The geographic sources of innovation: technological infrastructure and product innovation in the United States. Ann Assoc Am Geogr 84(2):210–229
Feldman M, Audretsch D (1999) Innovation in cities: science-base diversity, specialization and localized competition. J Eur Econ Rev 43:409–429
Fermeri D (2003) Marketing and management in high-technology sector: strategies and tactics in the commercial airplane industry. Praeger, Westport
Fidrmuc J (2000) Liberalization, democracy and economic performance during Transition. ZEI Economic Policy Papers, no. B05-2000, Centre for European Integration Studies, Bonn

Fingleton B (2000) Convergence: international comparisons based on a simultaneous equation model with regional effects. Int Rev Appl Econ 14(3):285–305, Taylor and Francis Journals

Fingleton B, McCombie JS (1998) Increasing returns and economic growth: some evidence for manufacturing from the European Union regions. Oxf Econ Pap 50(1):89–105

Fisher JC, Pry RH (1971) A simple substitution model of technological change. J Technol Forecast Soc Change 3(1):75–88

Fischer M, Varga A (2003) Spatial knowledge spillovers and university research: evidence from Austria. Ann Reg Sci 37(2):303–322, Springer

Fischer MT, Scherngell T, Jansenberger E (2005) The geography of knowledge spillovers between high-technology firms in Europe – evidence from a spatial interaction modelling perspective, ERSA conference paper 05(5), European Regional Science Association

Florida R, Gates G (2001) Technology and tolerance: the importance of diversity to high-technology growth. Centre on Urban and Metropolitan Policy, The Brookings Institution, Washington, DC

Foray D (1991) Morphological analysis, diffusion and lock-out of technologies: ferrous casting in France and the FRG. International Institute for Applied Systems Analysis, Laxenburg

Forman C, Goldfarb A, Greenstein SM (2005) Technology adoption in and out of major urban areas: when do internal firm resources matter most? National Bureau of Economic Research, Cambridge, MA

Francois J, Woerz J (2007) Producer services, manufacturing linkages, and trade. Tinbergen Institute Discussion Paper No. 045/2

Franzoni C, Lissoni F (2009) Academic entrepreneurs: critical issues and lessons for Europe. In: Varga A (ed) Universities, knowledge transfer and regional development: geography, entrepreneurship and policy. Edward Elgar, Cheltenham, pp 163–190

Freeman C (1982) The economics of industrial innovation. MIT, Cambridge, MA

Freeman C (1987) Technology policy and economic performance: lessons from Japan. Frances Pinter, London

Freeman PK (1994) Integrating environmental risk into corporate strategy. Risk Manage 41(7):54–59

Freeman C (2003) Policies for developing new technologies. SPRU Electronic Working Paper Series, No. 98, Sussex University

Freeman C, Soete L (eds) (1987) Technical change and full employment. Basil Blackwell, London

Freeman C, Soete L (1994) Work for all or mass unemployment. Pinter, London

Freeman C, Clark J, Soete L (1982) Unemployment and technical innovation. Pinter, London

Frenken, Ponds and Oort (2009) Innovation, spillovers, and university-industry collaboration: An extended knowledge production function approach, February 2009, http://ideas.repec.org/p/egu/wpaper/0903.html

Fritsch M, Werker C (1999) Systems of innovation in transition. In: Fritsch M, Brezinski H (eds) Innovation and technological change in Eastern Europe – pathways to industrial recovery. Edward Elgar, Cheltenham, pp 5–22

Fritsch M, Schwirten C (1999) Entreprise-university co-operation and the role of public research institutions in regional innovation systems. Ind Innov J 6:69–83

Fritsch M (2002) Measuring the quality of regional innovation systems: a knowledge production function approach. Int Rev Reg Sci 25:86–101

Fritsch M, Slavtchev V (2005) The role of regional knowledge sources for innovation. Working Paper 15/2005, Faculty of Economics and Business Administration, Technical University Bergakademie Freiberg

Fritsch M, Slavtchev V (2007) Universities and innovation in space. Ind Innov 14(2):201–218

Fritsch M, Medrano LF (2010) Regional knowledge and the emergence of an industry: laser systems production in West Germany, 1975–2005. Papers in Evolutionary Economic Geography # 10.16. www.econ.geo.uu.nl/peeg/peeg1016.pdf

Fuente A, Doménech R (2006) Human capital in growth regressions: how much difference does data quality make? J Eur Econ Assoc 4(1):1–36, MIT Press

Funke M, Niebuhr A (2001) Threshold effects and regional economic growth – evidence from West Germany. HWWA Discussion Papers 136, Hamburg Institute of International Economics (HWWA)

Gaczek W, Stryjakiewicz T (2008) Innowacyjna Wielkopolska. Ocena realizacji celów Regionalnej Strategii Innowacji oraz stanu systemu innowacji w Wielkopolsce, Urząd Marszałkowski Województwa Wielkopolskiego, Poznań www.wrpo.wielkopolskie.pl.
Gaczek WM (2009) Gospodarka oparta na wiedzy w regionach europejskich. Studia KPZK PAN, t. CXVIII. Komitet Przestrzennego Zagospodarowania Kraju PAN, Warszawa
Gadray J, Gallouj F, Weinstein O (1995) New modes of innovation. How services benefit industry. Int J Serv Ind Manag 6(3):4–16
Gaffard JL, Krafft J (2001) Telecommunications: understanding the dynamics of the organization of the industry, Université Nice Sophia Antipolis. http://hal.archives-ouvertes.fr/hal-00203655
Gaglio CM, Katz J (2001) The psychological basis of opportunity identification: entrepreneurial alertness. Small Bus Econ 16(2):95–111
Galaskiewicz J (1985) Interorganizational relations. Ann Rev Sociology J 11:281–304
Galbraith C, De Noble AF (1988) Location decisions by high technology firms: a comparison of firm size, industry type and institutional form. Entrep Theory Pract J 13:31–48, Baylor University
Gallouj F, Djellal F, Gallouj C (eds) (2007) The handbook of innovation and services. Edward Elgar, Northampton
Gans JS, Hsu DH, Stern S (2002) When does start-up innovation spur the gale of creative destruction? Rand J Econ 33(4):571–583
Gambardella A (2005) Patents and the division of innovative labor, In: Arora A, Merges R. Specialized supply firms, property rights, and firm boundaries, J Ind Corp Change 14 (6):1223–12233
Gaspar J, Glaeser EL (1998) Information technology and the future of cities. J Urban Econ 43(1):136–156, Elsevier
Gatignon H, Tushman M, Smith W, Anderson P (2002) A structural approach to assessing innovation: construct development of innovation locus, type and characteristics. Manag Sci 48(9):1103–1122
Georghiou L, Metcalfe S, Gibbons M, Evans J (1986) Post innovative performance. Macmillan, London
Geroski P (1989) Entry and the rate of innovation. Econ Innov New Technol 1:203–214
Gerschenkron A (1962) Economic backwardness in historical perspective. A book of essays. Harvard University Press, Cambridge
Gersick CJG (1991) Revolutionary change theories: a multilevel exploration of the punctuated equilibrium paradigm. Acad Manage Rev 16(1):10–36
Gertler MS (2003) A cultural economic geography of production: are we learning by doing? In: Anderson K, Domosh M, Pile S, Thrift N (eds) The handbook of cultural geography. Sage, London, pp 131–146
Gertler MS, Levitte Y (2003) Local nodes in global networks: the geography of knowledge flows in biotechnology innovation. Paper presented at the DRUID summer conference, Elsinore
Geuna A, Nesta LJJ (2006) University patenting and its effects on academic research: the emerging European evidence. J Res Policy 35:790–807, Elsevier
Geuna A, Rossi F (2011) Changes to university IPR regulations in Europe and the impact on academic patenting. Res Policy J 40(8):1068–1076, Elsevier
Giannetti M (2002) The effects of integration on regional disparities: convergence, divergence or both? Eur Econ Rev 46(3):539–567
Gibbons M et al (1994) The new production of knowledge: the dynamics of science and research in contemporary societies. Sage, London/Thousand Oaks
Giuliani E, Bell M (2005) The micro-determinants of meso-level learning and innovation: evidence from a Chilean wine cluster. Res Policy 34(1):47–68
Giuliani E (2007) The selective nature of knowledge networks in clusters: evidence from the wine industry. J Econ Geogr 7:139–168, Oxford Journals
Glaeser EL (1999) Urban primacy and politics. Harvard Institute of Economic Research, Cambridge, MA
Glaeser E, Kallal H, Scheinkman J, Shleifer A (1992) Growth in cities. J Polit Econ 100:1126–1152

Glaeser E, Scheinkman J, Shleifer A (1995) Economic growth in a cross-section of cities. J Monet Econ 36(1):117–143

Glaeser EW (2000) The new economics of urban and regional growth, In: Clark G, Gertler M, Feldman M (eds) The Oxford handbook of economic geography, Oxford University Press, Oxford, pp 83–98

Glasmeier A, Howland M (1994) Service-led rural development: definitions, theories, and empirical evidence. Int Reg Sci Rev 15:197–229

Gokhberg L (1994) Basic research in Russia: human resources and funding. Working Papers No 94060, International Institute for Applied Systems Analysis

Goldberg P, Amit K, Pavcnik N, Topalova P (2010) Imported intermediate inputs and domestic product growth: evidence from India. Q J Econ 125(4):1727, November 2010, Oxford University Press

Goto A, Suzuki K (1989) R&D capital, rate of return on R&D investment and spillover of R&D in Japanese manufacturing industries. Rev Econ Stat 71:555–564

Gottlieb PD, Fogarty M (2003a) Educational attainment and metropolitan growth. Econ Dev Q J 17:325–336, SAGE Pub

Gottlieb PD, Fogarty M (2003b) Educational attainment and metropolitan growth. J Econ Dev Q 17:325–336

Gort M, Klepper S (1982) Time paths in the diffusion of product innovations. Econ J 92(3):630–653, Wiley

Gorynia M, Jankowska B (2007) Koncepcja klasterów jako sposób regulacji zachowań podmiotów gospodarczych. Ekonomista 3:316

Gorzelak G (1996) The regional dimension of transformation in Central Europe. J. Kingsley, London

Gorzelak G (2009) Uniwersytet przedsiębiorczy. Forum Akademickie (1). http://forumakademickie.pl/szukaj/?q=gorzelak

Gorzelak G, Maier G, Petrakos G (2000) Integration and transition in Europe: the economic geography of interaction, Studies in the European economy. Routledge, New York

Gorzelak G, Bachtler J, Smętkowski M (2010) Regional development in central and eastern Europe. Development processes and policy challenges. Routledge, Oxon, p 338

Goulder LH, Schneider SH (1999) Induced technological change and the attractiveness of CO_2 abatement policies. Resour Energy Econ 21(3–4):211–253

Gourieroux C, Monfort A, Trognon A (1984) Pseudo maximum likelihood methods: applications to poisson models. J Econom 52(3):701–720

Gråsjö U (2008) University-educated labor, R&D and regional export performance. Int Reg Sci Rev 31(3):211–256

Grabher G (1997) Adaptation at the cost of adaptability. Restructuring the Eastern German Regional Economy. In: Grabher G, Stark D (eds) Restructuring networks: legacies, linkages, and localities in postsocialism. Oxford University Press, London

Graevenitz G, Wagner S, Harhoff D (2011) How to measure patent thickets – a novel approach. J Econ Lett 111(1):6–9, Elsevier

Greenan N, Guellec D (2001) Technological innovation and employment reallocation. J Labour 14 (4):547–590

Greunz L (2003) The technology gap and European regional growth dynamics. ULB Institutional Repository 2013/9485. Universite Libre de Bruxelles

Greunz L (2004) Interregional knowledge spillovers in Europe. ULB Institutional Repository 2013/9483. Universite Libre de Bruxelles

Greunz L (2005) Knowledge spillovers, innovation and social capital, ULB Institutional Repository 2013/9479, Universite Libre de Bruxelles

Griffith R, Redding S, Reenen JV (2003) R&D and absorptive capacity: theory and empirical evidence. Scand J Econ 105:99–118

Griffith R, Redding S, VanReenen J (2004) Mapping the two faces of R&D: productivity growth in a panel of OECD industries. Rev Econ Stat 86(4):883–895

Griffith R, Lee S, Van Reenen J (2011) Is distance dying at last? Falling home bias in fixed-effects models of patent citations. Quant Econ 2(2):211–249, ISSN 1759–7323
Griliches Z (1969) Capital-skill compementarity. Rev Econ Stat 51:465–468
Griliches Z (1979) Issues in assessing the contribution of research and development to productivity growth. Bell J Econ 10(1):92–116
Griliches Z (1980) R and D and the productivity slowdown. Am Econ Rev 70(2):343–348
Grimpe C, Patuelli R (2011) Regional knowledge production in nanomaterials: a spatial filtering approach. Ann Reg Sci 46(3):519–541, Springer
Grossetti M, Bes MP (2003) Dynamique de reseau et des cercles. Encastrements et decouplages. Journal of Revue d'Economie Industrielle, Numero special sur la morphogenese des reseaux 103(2):43–58
Grossman GM, Helpman E (1991) Trade, knowledge spillovers, and growth. Eur Econ Rev 35(2–3):517–526. Elsevier, April
Gruber H (1994) Learning and strategic product innovation: theory and evidence for the semiconductor industry. Amsterdam, North-Holland
Grubler A, Messner S (1998) Technological change and the timing of mitigation measures. Energy Econ 20(5–6):495–512
Grupp H (2000) Learning in a science-driven market: the case of lasers. Ind Corp Change 9:143–172
Guellec D, Van Pottelsberghe de la Potterie B (2004) From R&D to productivity growth: do the institutional settings and the source of funds of R&D matter? Oxford Bulletin of Economics and Statistics, Department of Economics, University of Oxford, 66(3):353–378
Guelleck D, Van Pottelsberghe de La Potterie B (1999) Does government support stimulate private R&D? OECD, Econ Stud J 29:95–122
Guellec D, van Pottelsberghe B (2001) The internationalisation of technology analyzed with patent data. Res Policy 30(8):1253–1266
Guellec D, van Pottelsberghe B (2004) From R&D to productivity growth: do the institutional settings and the source of funds of R&D matter? Oxf Bull Econ Stat 66(3):353–378
Guellec D, Van Pottelsberghe De La Potterie B (2003) The impact of public R&D expenditure on business R&D. Econ Innov New Technol 12(3):225–243
Gujarati D (1995) Basic econometrics, 3rd edn. McGraw-Hill International Editions, New York
Haffer M, Karaszewski W (eds) (2009) Czynniki wzrostu konkurencyjności przedsiębiorstw i regionów. Wydawnictwo UMK, Toruń
Hagedoorn J (2002) Inter-firm R&D partnerships: an overview of major trends and patterns since 1960. J Res Policy 31:477–492, Elsevier
Hagerstrand T (1952) The propagation of innovation waves, vol B-4. Lund studies in Geography, Lund
Hall BH, Ziedonis RH (2001) The patent paradox revisited: an empirical study of patenting in the US semiconductor industry, 1979–95. Rand J Econ 32(1):101–128
Hansen S (1993) Career development trends and issues in the United States. J Career Dev 20 (1):7–24
Harabi N (1997) Channels of R&D spillovers: an empirical investigation of Swiss firms. Technovation J 17(11/12):627–635
Harrison AE, Aitken BJ (1999) Do domestic firms benefit from direct foreign investment? Evidence from Venezuela. Am Econ Rev, Am Econ Assoc 89(3):605–618
Harrison R, Jaumandreu J, Mairesse J, Peters B (2008) Does innovation stimulate employment? A firm-level analysis using comparable micro-data from four European countries. NBER working papers 14216. National Bureau of Economic Research, Cambridge
Hart JA (1986) Politics of global competition in the semiconductor industry. Pac Focus 1 (2):93–119
Hassett KA, Hubbard RG (1996) Tax policy and investment, vol 5683, NBER Working NBER Working Paper. National Bureau of Economic Research, Cambridge

Hatzichronoglou T (1997) Revision of the high-technology sector and product classification. OECD STI working paper. OECD, Paris

Haupt R, Kloyer M, Lange M (2007) Patent indicators for the technology life cycle development. Res Policy 36(3):387–398

Hauser J, Tellis GJ, Griffin A (2006) Research on innovation: a review and Agenda for marketing science. Mark Sci 25(6):687–717

Häussler C, Zademach HM (2006) Cluster performance reconsidered: structure, linkages and paths in the German biotechnology industry, 1996–2003. Discussion paper series of SFB/TR 15 governance and the efficiency of economic systems, Free University of Berlin

Havas A (2006) Knowledge-intensive activities vs. high-tech sectors: traps and learning options for Central European policy-makers, In: Piech K, Radosevic S (eds) The knowledge- based economy in Central and East European Countries, Palgrave, pp 259–279

Head K, Ries J, Swenson D (1995) Agglomeration benefits and location choice: evidence from Japanese manufacturing investments in the United States. J Int Econ 38(3/4):223–247

Hecker DE (2005) High-technology employment: a NAICS-based update. Mon Labor Rev 128 (7):57–72

Henderson R, Cockburn I (1996) Scale, scope, and spillovers: determinants of research productivity in the pharmaceutical industry. Rand J Econ 27(1):32–59

Herz B, Vogel L (2003) Regional convergence in central and eastern Europe: evidence from a decade of transition. Bayreuth University economic discussion paper no. 13-03

Hicks D (2004) Real numbers Asian countries strengthen their research. Issues Sci Technol 20 (4):75–78

Hoecht A, Trott P (2006) Innovation risks of strategic outsourcing. Technovation 26 (5–6):672–681

Hoekman J, Frenken K, van Oort F (2010) The geography of collaborative knowledge production in Europe. Ann Reg Sci 43:721–738, Springer

Hollanders H (2006) Aerospace: scoping paper, research report commissioned by the European Commission in innovation watch – systematic, Maastricht

Hopkins M, Martin P, Nightingale P, Kraft A, Mahdi S (2007) The myth of the biotech revolution: an assessment of technological, clinical and organisational change. Res Policy 36:566–589

Hommen L, Doloreux D (2004) Bring back labour: a 'new' point of departure for the regional innovation approach. In: Flensburg P, Hörte SA, Karlsson K (eds) Knowledge spillovers and knowledge management in industrial clusters and industrial networks. Edward Elgar, London

Howitt P, Mayer-Foulkes D (2005) R&D, implementation, and stagnation: a schumpeterian theory of convergence clubs. J Money, Credit Bank 37(1):147–177, Blackwell Publishing

Inzelt A (2003) Foreign involvement in acquiring and producing new knowledge: the case of Hungary. In: Cantwell J, Molero J (eds) Multinational enterprises, innovative strategies and systems of innovation. Edward Elgar, Cheltenham/Northampton, pp 234–268

Irwin DA, Klenow PJ (1994) High-tech R & D subsidies: the effects of Sematech. Manuscript. University of Chicago, Chicago

Jacobs J (1969) The economy of cities. Random House, New York

Jaffe AB (1986) Technological opportunity and spillovers from R&D: evidence from firms' patents, profits, and market value. Am Econ Rev 76(5):984–1001

Jaffe AB (1989) Real effects of academic research. Am Econ Rev 79(5):957–970

Jaffe AB, Trajtenberg M (1996) Flows of knowledge from universities and federal labs. NBER Work Pap 5712:1–18

Jaffe AB, Trajtenberg M, Henderson R (1993) Geographic localization of knowledge flows as evidenced by patent citations. Q J Econ 108:577–598

Jaffe AB, Trajtenberg M (1998) International knowledge flows: evidence from patent citations. National Bureau of Economic Research, Cambridge

Jasinski AH (1994) R&D and innovation in Poland in the transition period. Econ Syst J 18:117–140

Jenelius E (2010) User inequity implications of road network vulnerability, Royal Institute of Technology. J Transport Land Use 2(3/4):57–73

Jindra B, Rojec M (2011) Foreign direct investment and knowledge spillovers in transition economies – is there a need for more policy coordination? J Knowl Econ (Special Issue: STI policy for the Enlarging European Economies), Springer

Jorgenson DW, Stiroh KJ (1995) Computers and growth. J Econ Innov New Technol 3 (3–4):295–316, Taylor&Francis

Jorgenson DW, Stiroh KJ (2000) Raising the speed limit: US economic growth in the information age, OECD Economics Department Working Papers 261, OECD Publishing

Jorgenson DW (2003) Information technology and the G7 economies, Harvard University, Mimeo Publisher. http://post.economics.harvard.edu/faculty/jorgenson/.

Jorgenson DW, Ho MS, Stiroh KJ (2008) A retrospective look at the U.S. productivity growth resurgence. J Econ Perspect 22(1):3–24, American Economic Association

Johansson B, Klaesson J (2001) Monopolistic competition, increasing returns, agglomeration, and transport costs. Ann Reg Sci 35(3):375–394, Springer

Johansson B, Klaesson J, Olsson M (2002) Time distances and labor market integration. Pap Reg Sci 81:305–327, Springer

Johansson B (2005) Parsing the menagerie of network and agglomeration externalities. In: Karlsson C, Johansson B, Stough R (eds) Industrial cluster and inter-firm networks. Edward Elgar, Cheltenham

Johnson DKN (2006) It's a small(er) world: the role of geography in biotechnology innovation. Colorado college working paper, Colorado

Johnson B, Lundvall B (2001) Why all this fuss about codified and tacit knowledge? Paper presented at the DRUID winter conference, Korsor

Johnson B, Lorenz E, Lundvall BA (2002a) Why all this fuss about codified and tacit knowledge? Ind Corp Change 11(2):245–262

Johnson DKN, Siripong N, Brown A (2002b) The demise of distance? The declining role of physical distance in knowledge transmission. Wellesley College Department of Economics working paper, Wellesley, Sept 2002

Jones-Evans D, Westhead P (1996) The high technology small firm sector in the UK. Int J Entrep Behav Res 2(1):15–35

Kaldor N (1960) Essays on value and distribution. Free Press, Glencoe

Kaldor N (1960) Esseys on economic growth and stability. Gerald Duckworth&Co, London

Karlsson C, Manduchi A (2001) Knowledge spillovers in a spatial context – a critical review and assessment. In: Fisher M, Frölich M (eds) Knowledge complexity and innovation systems. Springer, Germany

Kathawala Y, Zhang R, Shao J (2005) Global outsourcing and its impacts on organizations: problems and issues. Int J Serv Oper Manag 1(2):185–202

Katsoulacos Y (1986) The employment effect of technical change. A theoretical study of new technology and the labour market. Wheatsheaf, Brighton

Katz L, Krueger A (1997) Computing inequality: have computers changed the labor market? NBER Working Paper No. 5956, National Bureau of Economic Research, Cambridge

Katz N, Du Preez N (2008) The role of knowledge management in supporting a radical innovation project. In: Bernard A, Tichkiewitch S (eds) Methods and tools for effective knowledge life-cycle management. Springer, Amsterdam, pp 331–346

Kauffman RJ, Kumar A (2006) A combined scale-and-scope theory of IT industry cluster growth. In: The 2006 IEEE international society for engineering management conference proceedings, IEEE computing society, Los Alamitos

Kaushalesh L, Dunnewijk T (2008) Entrepreneurship and innovation strategies in ICT SMEs in enlarged Europe (EU-25), vol 16, UNU-MERIT working paper series. UNU-MERIT, Maastricht

Keeble D, Wilkinson F (2000) High-technology clusters, networking, and collective learning in Europe. Aldershot, Hampshire

Kelejian HH, Prucha IR, Yuzefovich E (2004) Instrumental variable estimation of a spatial autoregressive model with autoregressive disturbances: large and small sample results. In: LeSage J, Pace K (eds) Advances in econometrics: spatial and spatiotemporal econometrics. Elsevier, New York, pp 163–198

Keller W (1997) Are international R&D spillovers trade-related? Analyzing spillovers among randomly matched trade partners. Eur Econ Rev 42(8):1469–1481

Keller W (2001) International technology diffusion. NBER Working Papers 8573, National Bureau of Economic Research

Kelly M, Hageman A (1999) Marshallian externalities on innovation. J Econ Growth 4:39–54, Springer

Kendrick JW (1981) International comparisons of recent productivity trends. In: Fellner W (ed) Essays in contemporary economic problems. American Enterprise Institute, Washington, DC

Ketelhöhn NW (2006) The role of clusters as sources of dynamic externalities in the US semiconductor industry. J Econ Geogr 6:679–699, Oxford

Khan TS (2006) Productivity growth, technological convergence, R&D, trade, and labor markets: evidence from the French manufacturing sector, IMF Working Paper 06/230

Klein L (1974) A textbook of econometrics, 2nd edn. Prentice-Hall, Englewood Cliffs

Kline S, Rosenberg N (1986) An overview of innovation. In: Landau R, Rosenberg N (eds) The positive sum strategy: harnessing technology for economic growth. National Academy Press, Washington, DC

Kline S, Rosenberg N (1987) An overview of Innovation. In: Landau R, Rosenberg N (eds) The positive sum strategy: harnessing technology for economic growth. National Academy Press, Washington, DC

Knell M, Rojec M (2007) The economics of knowledge and knowledge accumulation: a literature survey. Mimeo Publisher, Prague

Kneller R (1999) University-industry cooperaiton in biomedical R&D in Japan and the United States: implications for biomedical industries. In: Branscomb L et al (eds) Industrializing knowledge: university-industry linkage in Japan and the United States. MIT Press, Cambridge, MA

Kneller R (2001) Technology transfer: a review for biomedical researchers. Clin Cancer Res J 7:761–774

Knudsen B, Florida R, Gates G, Stolarick K (2007) Urban density, creativity and innovation. Working paper, The Martin Prosperity Institute, University of Toronto

Kokko A, Tansini R, Zejan M (1996) Local technological capability and spillovers from FDI in the Uruguayan manufacturing sector. J Dev Stud 34:602–611

Kolasa M (2005) What drives productivity growth in the new EU member States? The case of Poland. European Central Bank working paper series, No. 486

Kołodko GW (2008) Truth, errors, and lies. Politics and economics in a volatile world. Colombia University Press, New York

Kołodko GW (2011) Truth, errors and lies. Politics and economics in a volatile world. Columbia University Press, New York, p 106

Kołodko GW, Piatkowski M (2002) Nowa gospodarka i stare problemy. Perspektywy szybkiego wzrostu w krajach posocjalistycznych. Wydawnictwo WSPiZ, Warszawa

Kondo M (2005) Dynamic analyses on the relation between R&D and patent application in Japan. J Sci Policy Res Manag 10(3–4):193–204

Kosiedowski W (2005) Regiony Europy Środkowej i Wschodniej wobec globalizacji i integracji międzynarodowej. Wydawnictwo UMK, Włocławek

Kozłowski J (2010) Statystyka nauki, techniki I innowacji w krajach UE i OECD. Stan i problemy rozwoju. Departament Strategii MNiSW, Warszawa

Koźmiński AK (2001) Jak tworzyć gospodarkę opartą na wiedzy. w: Strategia rozwoju Polski u progu XXI wieku, Komitet Polska 2000 Plus, Warszawa, s 32–39

Koźmiński AK (2005) Building up competitive advantage in the new Europe: the role of higher education. In: Aleksandre W, Gasparski WW (eds) French and other perspectives in

praxiology. Praxiology: the international annual of practical philosophy and methodology. New Transaction, New Brunswick/London, pp 191–212
Koźmiński A, Yip G (2000) Strategies for central and eastern Europe. Palgrave Macmillan, New York
Krammer MS (2009) Drivers of national innovation in transition: evidence from a panel of Eastern European countries. Res Policy J 38:845–860, Elsevier
Kremer M, Onatski A, Stock JH (2001) Searching for prosperity. National Bureau of Economic Research, Cambridge, MA
Kruckenberg L, Brandes A, Ahrweiler P (2008) R&D governance rules of the EU framework programmes: the EURuleD archive. NEMO (Supported by the EU FP6-NEST-Adventure Programme Contract n° 028875) Working Paper 6
Krugman P (1991) Geography and trade. Leuven University Press/The MIT, Leuven
Kukliński A (2001) The role of universities in stimulating regional development and educating global elites. High Educ Eur 26(3):437–445
Lacity MC, Willcocks LP, Feeny DF (1995) IT outsourcing: maximize flexibility and control. Harv Bus Rev 73(3):84–93
Lai E, Riezman R, Wang P (2009) Outsourcing of innovation. Econ Theory J 38(3):485–515, Elsevier
Lamoreaux NR, Sokoloff KL (1996) Long-term change in the organization of inventive activity. Proc Natl Acad Sci USA 93(23):12686–12692
Lane PJ, Lubatkin M (1998) Relative absorptive capacity and interorganizational learning. Strateg Manag J 19:461–477, Wiley-Blackwell
Langlois RN (1990) Creating external capabilities: innovation and vertical disintegration in the microcomputer industry. Bus Econ Hist 19(1):93–102
Laursen K, Salter A (2006) Open for innovation: the role of openness in explaining innovation performance among U.K. manufacturing firms. Strateg Manag J 27(2):131–150
Lawton-Smith H, Bagchi-Sen S (2004) Innovation geographies: international perspectives on research, product development, and commercialisation of biotechnologies. Environ Plann C Govern Policy 22(2):159–160
Leamer EE, Storper M (2001) The economic geography of the internet age. J Int Bus Stud 32 (4):641–665
Le Gallo J, Ertur C (2003) Exploratory spatial data analysis of the distribution of regional per capita GDP in Europe, 1980–1995. J Pap Reg Sci 82(2):175–201
Ledesma ML (1999) Accumulation, innovation and catching-up: an extended cumulative growth model. Journal of Studies in Economics No 9906, Department of Economics, University of Kent
Lee JW (2008) Comment on rapid economic growth: contributing factors and challenges ahead. Asian Econ Policy Rev 3(2):207–208
Leiblein MJ, Miller DJ (2003) An empirical examination of transaction- and firm-level influences on the vertical boundaries of the firm. Strateg Manag J 24(9):839–859
Leiblein MJ, Reuer J (2004) An analysis of the effects of firm capabilities and international collaboration on the foreign sales of semiconductor firms. J Bus Ventur 19(1):285–307
Leonard-Barton D (1992) Core capabilities and core rigidities. Strateg Manag J 13:111–126, Summer special issue
Lesage V, Gosselin JF, Hammill MO, Kingsley MCS, Lawson JW (2007) Ecologically and Biologically Significant Areas (EBSAs) in the Estuary and Gulf of St. Lawrence - A marine mammal perspective. Research Document. Can Sci Adv Sec 2007/046:1–94
Levin RC, Klevorick AK, Nelson RR, Winter SG, Gilbert R, Griliches Z (1987) Appropriating the returns from industrial research and development. Brook Pap Econ Act Spec Issue Microecon 3:783–831
Levinthal DA, March JG (1993) The myopia of learning. Wharton School, Philadelphia
Lengyel B, Cadil V (2009) Innovation policy challenges in transition countries: foreign business R&D in the Czech republic and Hungary. Transition Stud Rev J 16:174–188, Springer

Leydesdorff L, Etzkowitz H (eds) (1997) Universities in the global economy: a triple helix of University-Industry-Government Relations. Cassell Academic, London

Lichtenberg FR (1993) R&D investment and international productivity differences. NBER Working Papers 4161, National Bureau of Economic Research

Lim U (2003) Knowledge spillovers and regional economic growth and convergence in U.S. metropolitan areas. Paper read at 18th annual meeting of the Pacific Regional Science Conference Organization, Acapulco. www.sussex.ac.uk/Units/spru/events/KP_Conf_03/.../Sonn_Storper.pdf

Link AN, Siegel DS (2005) University-based technology initiatives: quantitative and qualitative evidence. Res Policy J 34(3):253–257, Elsevier

Lockett A, Wright M (2005) Resources, capabilities, risk capital and the creation of university spin-out companies, technology transfer and universities' spin-out strategies. Res Policy J 34 (7):1043–1057, Elsevier

Lopez-Bazo E, Vaya E, Mora AJ, Surinach J (1999) Regional economic dynamics and convergence in the European union. J Ann Reg Stud 33:343–370

Los B, Verspagen B (2006) The evolution of productivity gaps and specialization patterns. Metroeconomica 57(4):464–493

Lucas RE (1988) On the mechanics of economic development. J Monet Econ 22:3–42

Lucas RE (1993) Making a miracle. J Econom 61(2):251–272, Econometric Society

Lundvall B-A (1992) National systems of innovation: towards a theory of innovation and interactive learning. Pinter Publishers, London

Lundvall B-A, Borrás S (1998) The globalising learning economy: implications for innovation policy. European Commission, Brussels

Lundquist K-J, Olander L-O (1999) Firms, regions and competitiveness. A broad-brush approach. Geografiska Annaler Series 81B:145–163

Macher JT, Mowery DC, Hodges DA (1999) Semiconductors. In: Mowery DC (ed) U.S. industry in 2000: studies in competitive performance. National Academy Press, Washington, DC

MacKinnon DP, Lockwood CM, Hoffman JM, West SG, Sheets V (2002) A comparison of methods to test mediation and other intervening variable effects. Psychol Methods 7:83–104

Maddison A (1982) Phases of capitalist development. Oxford University Press, Oxford

Maddison A (1991) Dynamic forces in capitalist development. Oxford University Press, Oxford

Maggioni MA, Uberti TE (2007) International networks of knowledge flows: an econometric analysis. In: Frenken K (ed) Applied evolutionary economics and economic geography. Edward Elgar, Cheltenham, UK

Maggioni MA, Nosvelli M, Uberti TE (2007a) Space vs. networks in the geography of innovation: a European analysis. Pap Reg Sci 86(3):471–493

Maggioni MA, Nosvelli M, Uberti TE (2007) Space vs. networks in the geography of innovation: a European analysis. Pap Reg Sci 86(3):471–493

Makun P, MacPherson AD (1997) Externally- assisted product innovation in the manufacturing sector: the role of location, in-house R&D and outside technical support. Reg Stud 31(7):659–688

Malecki E (1985) Industrial location and corporate organization in high-technology industries. Econ Geogr J 61:345–367, Oxford Journals

Malecki EJ (1991) Technology and economic development: the dynamics of local, regional and national change. Longman, London

Malerba F (1985) The semiconductor business – the economics of rapid growth and decline. University of Wisconsin Press/Frances Pinter, Madison/London

Malerba F (1992) Learning by firms and incremental technical change. Econ J 102(413):845–859

Malerba F, Orsenigo L (1994) Schumpeterian patterns of innovation. Camb J Econ 19:47–66

Manley K (2002) The systems approach to innovation studies. Aust J Info Syst 9(2):94–102

Markman G, Phan P, Balkin D, Gianiodis P (2005) Entrepreneurship and university-based technology transfer. J Bus Venturing 20(2):241–263, Elsevier

Marks K (1973) Grundrisse. Foundations of the critique of political economy. Vintage Books, New York

Markusen A (1994) Studying regions by studying firms. J Prof Geogr 46:477–490
Marshall A (1890) Principles of economics. Macmillan, London
Marshall A (1920) Principles of economics; an introductory volume. Macmillan, London
Marshall JN, Damesick P, Wood P (1987) Understanding the location and role of producer services in the United Kingdom. Environ Plann A 19:575–596
Marx K (1973) Grundrisse. Penguin Classics, London
Maurseth PB, Verspagen B (2002) Knowledge spillovers in Europe: a patent citations analysis. Scand J Econ 104(4):531–545
McEvily B, Zaheer A (1999) Bridging ties: a source of firm heterogeneity in competitive capabilities. Strateg Manag J 10(12):1133–1156
McKelvey M, Alm H, Riccaboni M (2003) Does co-location matter for formal knowledge collaboration in the Swedish biotechnology-pharmaceutical sector? Res Policy 32:483–501
McKelvie A (2007) Innovation in new firms: examining the role of knowledge and growth willingness. JIBS, Jonkoping
McNiven C (2007) Overview and discussion of the results of the Pilot Survey on Nanotechnology in Canada. Science, Innovation and Electronic Information Division (SIEID). Working Paper No. 005
Menzel MP, Fornahl D (2007) Cluster life cycles - Dimensions and Rationales of Cluster Development, Max- Paper at the DRUID - DIME Academy Winter 2007 PhD Conference on "Geography, Innovation and Industrial Dynamics" in Skoerping, 25–27 January
Metters R, Verma R (2008) History of offshoring knowledge services. J Oper Manag 26:141–147, Elsevier
Meriküll J (2008) The impact of innovation on employment: firm- and industry-level evidence from Estonia. Bank of Estonia working paper no. 1/2008, Tallinn
Minshall T, Druilhe C, Probert D (2004) The evolution of "third mission" activities at the University of Cambridge: balancing strategic and operational considerations. Paper presented at the 12th high tech small firms conference, University of Twente, The Netherlands
Moomaw RL (1981) Productivity and city size: a review of the evidence. Q J Econ 96:675–688
Mora T (2008) Factors conditioning the formation of European regional convergence clubs. Ann Reg Sci 42:911–927
Moreno Y, Pastor-Satorras R, Vazquez A, Vespignani A (2003) Critical load and congestion instabilities in scale-free networks. Europhys Lett 62(2):292–298
Moreno A, Cacho I, Canals M, Grimalt JO, Sanchez-Vidal A (2003b) Millennialscale variability in the productivity signal from the Alboran Sea record, Western Mediterranean Sea. Palaeogeogr Palaeoclimatol Palaeoecol 211:205–219
Moreno R, Paci R, Usai S (2005) Spatial spillovers and innovation activity in European regions. Environ Plann A 37:1793–1812
Morgan K (1997) Thelearning region: institutions, innovation and regional renewal. Reg Stud 31 (5):491–503
Moriarty RT, Kosnik JT (1989) High-tech marketing: concepts, continuity, and change. Sloan Manage Rev 30(4):7–17
Morrison CJ, Siegel D (1997) Automation or openness? Technology and trade impacts on costs and labor composition in the food system. Department of Resource Economics Regional Research Project 965, University of Massachusetts
Morrison A, Rabellotti R (2009) Knowledge and information networks in an Italian wine cluster. Taylor&Francis: Eur Plann Stud 17(7):983–1006
Mowery D, Rosenberg N (1979) The influence of market demand upon innovation: a critical review of some recent empirical studies. Res Policy 8(2):102–153
Mowery D, Rosenberg N (1989) Technology and the pursuit of economic growth. Cambridge University Press, Cambridge
Mukkala K (2011) Essays on regional development and labor mobility in a knowledge-based economy, Jyväskylä studies in business and economics. University of Jyvaskyla, Jyvaskyla
Murphy KM, Riddell WC, Romer PM (1998) Wages, skills, and technology in the United States and Canada. National Bureau of Economic Research, Cambridge, MA

Murray F (2004) The role of academic inventors in entrepreneurial firms: sharing the laboratory life. Res Policy 33(4):643–659

Murray F, Stern S (2007) When ideas are not free: the impact of patents on scientific research. In: Jaffe AB, Lerner J, Stern S (eds) Innovation policy and the economy, vol 7. MIT Press, Cambridge, MA, pp 33–69

Mytelka L, Farinelli F (2000) Local clusters, innovation systems and sustained competitiveness. Discussion papers 5. United Nations University, Institute for New Technologies

Narula R (2004) Understanding absorptive capacities in an "innovation systems" context consequences for economic and employment growth. DRUID Working Papers 04-02, Copenhagen Business School, Aalborg University

Neffke F, Svensson Henning M (2008) Revealed relatedness: mapping industry space. Papers in Evolutionary Economic Geography 08(19), Urban and Regional Research Centre, Utrecht

Neffke F, Henning S, Boschma RA, Lundquist K, Olander L (2009) The dynamics of agglomeration externalities along the life cycle of industries. Regional Studies Journal, Taylor & Francis

Nelson RR (ed) (1993) National innovation systems: a comparative analysis. Oxford University Press, New York

Nelson RR, Winter SG (1982) An evolutionary theory of economic change. Belknap Press, Cambridge, MA

Nelson R (2004) The market economy and the scientific commons. Res Policy 33:455–471

Nerkar A, Roberts PW (2004) Technological and product-market experience and the success of new product introductions in the pharmaceutical industry. Strateg Manag J 25(8–9):779–799

Nesvetailov GA, Artuykhin MI (1995) Brain drain in Belarus. J Sci Stud 8:73–77, SAGE Pub

Nonaka I, Takeuchi H (1995) The knowledge-creating company: how Japaneese companies create the dynamics of innovation. Oxford University Press, New York

Nonaka I, Toyama R, Konno N (2000) SECI, Ba, and leadership: a unified model of dynamic knowledge creation. Long Range Plann 33:5–34

Nooteboom B (2000) Learning and innovation in organizations and economies. Oxford University Press, Oxford

Nusser M, Hüsing B, Wydra S (2007) Potenzialanalyse der industriellen, weißen Biotechnologie. Studie im Auftrag des Bundesministeriums für Bildung und Forschung (BMBF), Karlsruhe

OECD (2003) Genetic inventions, IPRs and licensing practices: evidence and policies. OECD, Paris

OECD (2012) Patent statistics. http://www.oecd.org/innovation/inno/oecdworkonpatentstatistics.htm

Okoń-Horodyńska E (2007) Building the polish knowledge based economy. Basic dilemmas in the time of system transition. In: Banse G (ed) Technological and environmental policy. Studies in eastern European thought, vol 5. Sigma, Berlin, pp 275–303

Okoń-Horodyńska E (2008) Edukacja dla innowacji. J Nauka i Szkolnictwo Wyższe, 1/31/2008: 34–54

Olechnicka A (2012) *Potencjał* Nauki a Innowacyjnosc Regionow. Wydawnictwo Scholar, Warszawa, p 171

Orlando MJ, Verba M (2005) Do only big cities innovate? Technological maturity and the location of innovation, Economic Review. Federal Reserve Bank of Kansas City, Second Quarter, pp 31–57

Ortega JL, Aguillo IF (2010) Shaping the European research collaboration in the 6th framework programme health thematic area through network analysis. J Scientometrics 85:377–386

Ostergaard CR (2008) Knowledge flows through social networks in a cluster: comparing University and industry links. In: Proceedings of the 5th international EMAEE conference, Metropolitan University, Manchester. www.cjournal.cz/files/56.pdf

Ota M, Fujita M (1993) Communication technologies and spatial organization of multi-unit firms in metropolitan areas. Reg Sci Urban Econ 23:695–729, Elsevier

Ouwersloot H, Rietveld P (2000) The geography of R&D: tobit analysis and a Bayesian approach to mapping R&D activities in the Netherlands. Tinbergen Institute, Amsterdam

Owen-Smith J, Powell WW (2004) Knowledge networks as channels and conduuits: the effects of spillovers in the Boston biotechnology community. Organ Sci 15(1):5–21
Owen-Smith J, Riccaboni M, Pammolli F, Powell WW (2002) A comparison of U.S. and European university-industry relations in the life sciences. Manag Sci 48(1):24–43
Paci R, Usai S (1999) The role of specialisation and diversity externalities in the agglomeration of innovative activities, Working Paper CRENoS 15, Centre for North South Economic Research, University of Cagliari and Sassari, Sardinia
Paci R, Usai S (2000) Externalities, knowledge spillovers and the spatial distribution of innovation. ERSA conference papers 104, European Regional Science Association
Paci R, Pigliaru F (2001) Technological diffusion, spatial spillovers and regional convergence in Europe. Working Paper CRENoS 200101, Centre for North South Economic Research, University of Cagliari and Sassari, Sardinia
Paci R, Pigliaru F (2002) Technological diffusion, spatial spillovers and regional convergence in Europe. In: Cuadrado-Roura JR, Parellada M (eds) Regional convergence in the European union. Springer, Berlin, pp 273–292
Parent O, Riou S (2005) Bayesian analysis of knowledge spillovers in European regions. J Reg Sci 45(4):747–775
Parente SL, Prescott EC (2000) Barriers to riches. MIT Press, Cambridge
Patel P, Pavitt K (1991) Large firms in the production of the world's technology – an important case of non-globalization. J Int Bus Stud 22(1):1–21
Patel P, Vega M (1999) Patterns of internationalisation of corporate technology: location vs. home country advantages. J Res Policy 28:145–155
Pavitt K (1998) Technologies, products and organization in the innovating firm: what Adam Smith tells us and Joseph Schumpeter doesn't. Ind Corp Change 7:433–457
Pennings JM (1988) Technological innovations in manufacturing organizations. Int Stud Manag Organ xvii:68–89
Peterson J, Sharp M (1998) Technology policy in the European Union. St. Martin's, New York
Petit P, Soete L (1998) Globalization in search of a future: the contemporary challenge to national policies. CEPREMAP Research Center, Working Papers, No 9819, Paris. http://www.cepremap.ens.fr/version/docume/index_en.php?choix=DOCWEB
Petruzzelli AM (2009) University-industry R&D collaborations: a joint-patents analysis. Paper to be presented at the summer conference on June 17–19 in Copenhagen Business School. http://www2.druid.dk/conferences/viewpaper.php?id=5731&cf=32
Pianta M (2005) Innovation and employment. In: Fagerberg J, Mowery D, Nelson R (eds) Handbook of innovation, Chap. 22. Oxford University Press, Oxford/New York, pp 568–598
Pianta M (2005b) Innovation and employment. In: Fagerberg J, Mowery D, Nelson R (eds) The oxford handbook of innovation. Oxford University Press, Oxford, pp 568–598
Piatkowski M (2003) The 'new economy' and catching-up potential of transition economies. Prague economic papers 1/2003, pp 37–56
Piergiovanni R, Santarelli E (2001) Patents and the geographic localization of R&D spillovers in French manufacturing. Reg Stud J 35:697–702, Taylor&Francis
Ponds R, Van Oort F, Frenken K (2007) The geographical and institutional proximity of research collaboration. Reg Sci J 86(3):423–443, Taylor&Francis
Pons-Novell J, Viladecans-Marsal E (1999) Kaldor's laws and spatial dependence: evidence for the European regions. Reg Stud 33:443–451
Porter ME (1990, 1998) The competitive advantage of nations. Free Press, New York
Porter ME (1998) The competitive advantage of nations. Free Press, New York
Porter ME (2000) Location, competition, and economic development: local clusters in a global economy. Econ Dev Q 14(1):15–34
Porter M (2003) Clusters and regional competitiveness: recent learnings. Paper presented at the international conference on technology clusters, Montreal, 7 November 2003 www.isc.hbs.edu/pdf/Montreal_Cluster_Conference_2003.11.07_ckrb.pdf

Porter K, Whittington KB, Powell WW (2005) The institutional embeddedness of high-tech regions: relational foundations of the Boston biotechnology community. In: Breschi S, Malerba F (eds) Clusters, networks, and innovation. Oxford University Press, Oxford, pp 261–296

Porter M (2008) On competition, updated and expanded edition. Harvard Business School Publishing, Cambridge

Powell W (1998) Learning from collaboration: knowledge and networks in the biotechnology and pharmaceutical industries. Calif Manage Rev 40(3):228–240

Powell W, Koput KW, Smith-Doerr L (1996) Interorganizational collaboration and the locus of innovation: networks of learning in biotechnology. Adm Sci Q 41(1):116–145

Prevenzer M (1997) The dynamics of industrial clustering in biotechnology. J Small Bus Econ 9:255–271, Springer

Protogerou A, Caloghirou Y, Siokas E (2010) Policy-driven collaborative research networks in Europe. Econ Innov New Technol 9(4)

Protogerou A, Caloghirou Y, Lioukas S (2011) Dynamic capabilities and their indirect impact on firm performance. J Ind Corp Change, 1–33

Putnam J, Evenson RE (1994) Inter-sectoral technology flows: estimates from a patent concordance with an application to Italy. Yale University, Mimeo Publisher

Pyka A, Küppers G (2002) Innovation networks. Edward Elgar, Cheltenham

Quinn JB (2000) Outsourcing innovation: the new engine of growth. Sloan Manag Rev J 41 (4):13–28

Radosevic S (2000) Divergence or convergence in R&D and innovation between "East" and "West". In: Fritsch M, Brezinski H (eds) Innovation and technical change in eastern Europe. Pathways to industrial recovery. Edward Elgar, Cheltenham

Radosevic S (2002) The electronics industry in central and eastern Europe: an emerging production location in the alignment of networks perspective. Working papers 21. School of Slavonic and East European Studies, University College London

Radosevic S, Auriol L (1999) Patterns of restructuring in research, development and innovation activities in central and eastern European countries: analysis based on S&T indicators. Res Policy 28:351–376

Rajasalu T, Laur A (2003) Contribution of the new economy to Estonia's economic growth and convergence with the European economy. TIGER working paper series no. 41, pp 1–17. www.tiger.edu.pl/publikacje/TWPNo41.pdf

Rapacki R, Prochniak M (2009) Real beta and sigma convergence in 27 transition countries, 1990–2005. Post-Communist Econ J 21(3):307–326, Taylor and Francis

Rasheed A, Manarvi IA (2008) A framework of technology diffusion in Aircraft Manufacturing Industry Environment. Lecture Notes in Engineering and Computer Science 01/2008. www.researchgate.net/publication/44261831_A_Framework_of_Technology_Diffusion_in_Aircraft_Manufacturing_Industry_Environment

Rey S, Montouri B (1999) US regional income convergence: a spatial econometric perspective. J Reg Stud 33:143–156

Rigby DL, Essletzbichler J (1997) Evolution, process variety, and regional trajectories of technological change in US manufacturing. Econ Geogr 73(3):269–284

Rodriguez-Pose A, Crescenzi R (2006) R&D, spillovers, innovation systems and the genesis of regional growth in Europe. ERSA conference papers. European Regional Science Association, p 371. http://www.ersa.org

Roediger-Schluga T, Barber MJ (2006) The structure of R&D collaboration networks in the European Framework Programmes. UNU-MERIT working paper, 2006–036. http://arno.unimaas.nl/show.cgi?fid=6621

Romer PM (1986) Increasing returns and long-run growth. J Polit Econ 94(5):1002–1037

Romer PM (1990) Endogenous technological change. J Polit Econ 98(5):71–102

Romer PM (1993) Economic growth. Canadian Institute for Advanced Research, Toronto

Rõõm M (2002) Unemployment and labour mobility in Estonia. Analysis using duration models, Working Papers, 7. Bank of Estonia, Tallinn

Ropohl G (1997) Knowledge types in technology. Int J Technol Des Educ 7(1–2):65–72
Rosenberg N (1972) Factors affecting the diffusion of technology. Explor Econ Hist 10(1):3–33
Rosenthal SS, Strange WC (2003) Geography, industrial organization, and agglomeration. Rev Econ Stat 85(2):377–393, MIT Press
Rugraff E (2008) Are the FDI policies of the Central European countries efficient? Post-Communist Econ 20(3):303–316
Runiewicz M (2005) The impact of foreign direct investments on the export competitiveness of the Baltic States in the European single market. Kriventa Publishing, Vilnius
Runiewicz-Wardyn M (2005) The impact of foreign direct investments on the export competitiveness of the Baltic States in the European single market. Kriventa, Vilnius
Runiewicz-Wardyn M (2006) Znaczenie ICT w generowaniu zdolnosci konkurencyjnej regionu. In: Runiewicz-Wardyn M (ed) Konkurencyjność regionów. Rola technologii informatyczno-telekomunikacyjnych. Wyd. ALK, Warsaw
Runiewicz-Wardyn M (2007) Patent Europejski. Perspektywa polskich jednostek badawczo-rozwojowych. Master Bus Adm 3(86):42–56
Runiewicz-Wardyn M (2008a) Evolution of factors and measures of international competitivenes. The increasing role of knowledge. In: Runiewicz-Wardyn M (ed) Knowledge-based economy as factor of competitiveness and economic growth, Wyd. Akademickie i Profesjonalne, Warsaw
Runiewicz-Wardyn M (2008b) Towards building and eLearning environment in Poland. Wyd. Akademickie i Profesjonalne, Warsaw
Runiewicz-Wardyn M (2010) Estońska droga ku gospodarce opartej na wiedzy. In: Gołębiowski J (ed) Państwa Bałtyckie w zintegrowanej Europie, Wyd. WSCL, Warszawa
Runiewicz-Wardyn M (2011) Dynamika przemysłów wysokich technologii a rozwój klastrów wiedzy na przykładzie klastra Route 128. Master Bus Adm J 4:47–68
Runiewicz-Wardyn M (2012a) The innovation process of European regions: from absorption to knowledge creation capacities of European regions. In: Olechnicka A, Capello R, Gorzelak G (eds) Universities, cities and regions. Loci for knowledge and innovation creation. Routledge, London
Runiewicz-Wardyn M (2012b) Tworzenie i upowszechnianie wiedzy w branży biotechnologicznej na przykładzie regionów Flandrii i Walonii w Belgii. Master Bus Adm J 5(118):100–113
Ruttan VW (2006) Social science knowledge and induced institutional innovation: an institutional design perspective. J Inst Econ 2(3):249–272
Saggi K (2002) Trade, foreign direct investment, and international technology transfer: a survey. World Bank Res Obs 17(2):191–235, Oxford University Press
Sakakibara M, Brandstetter L (2003) Measuring the impact of US research consortia. Manage Decis Econ J 24:51–69, Wiley Pub
Sakurai N, Ioannidis E, Papaconstantinou G (1996) The impact of R&D and technology diffusion on productivity growth: evidence for 10 OECD countries in 1970s and 1980s, vol 1996/2, STI working paper. OECD, Paris
Sapsed J, Grantham A, DeFillippi R (2007) A bridge over troubled waters: bridging organizations and entrepreneurial opportunities in emerging sectors. Res Policy 36:1314–1334
Sargossi S, de la Potterie B v P (2003) What patent data reveal about universities: the case of Belgium. J Technol Transfer 28:47–51
Savy M (1998) TIC et territoire: le paradoxe de localiosation, published by Laboratoire d'Economie des Transports Institut des Sciences de l'Homme, Les Cahiers Scientifiques du Transport, No. 33:129–146
Saxenian AL (1994) Regional advantage: culture and competition in Silicon Valley and Route 128. Harvard University Press, Cambridge, MA
Saxenian AL (2006) The new argonauts: regional advantage in a global economy. Harvard University Press, Cambridge
Scherer FM (1982) Interindustry technology flows and productivity growth. Rev Econ Stat 64:627–634

Scherer FM (1999) New perspectives on economic growth and technological innovation. Brookings Institution Press, Washington, DC

Scherngell T, Barber M (2009) Spatial interaction modelling of cross-region R&D collaborations: empirical evidence from the 5th EU framework programme. Paper Reg Sci, Wiley Blackwell 88(3):531–546

Scherngell T, Barber M (2010) Distinct spatial characteristics of industrial and public research collaborations: evidence from the 5th EU framework programme. Ann Reg Sci 46:247–266

Scherngell T, Lata R (2011) The geographical dynamics of R&D networks in Europe: a regional perspective using framework programme data. Paper presented at the 2011 EAEPE congress, Vienna, 27–29 Oct 2011

Schumpeter JA (1942) Capitalism, socialism and democracy. Harper, New York/London

Seifoddin A et al (2008) Toward evolutionary innovation theory. Int J Eng 19(1):43–55

Senator Lieberman JI (2004) Offshore outsourcing and America's competitive edge: losing out in the high technology R&D and services sectors. United States Senate, Washington, DC

Sepp L, Varblane U (2009) How to improve the supportive role of Estonian innovation system toward launching new products by high technology companies? Publishing House of University of Tartu, pp 357–372

Shane S (2000) Prior knowledge and discovery of entrepreneurial opportunities. Organ Sci 11 (4):448–469

Shane S (2005) The effects of patent litigation on university licensing efforts. NBER Reporter (2)

Shapiro C (2002) Competition policy and innovation. STI working papers 2002/11. OECD. www.oecd.org/sti

Sharma A, Iyer GR, Evanschitzky H (2008) Personal selling of high-technology products: the solution-selling imperative. J Relat Mark 7(3):287–308

Sheehan J, Guellec D, Martinez C (2003) Business patenting and licensing: results from the OECD/BIAC survey. In: Patents innovation and economic performance, proceedings of the OECD conference on IPR, innovation and economic performance, 28–29 Aug 2003. OECD. http://www.oecd.org/sti/sci-tech/patentsinnovationandeconomicperformanceconferenceproceedings.htm

Simon C (1998) Human capital and metropolitan employment growth. J Urban Econ 43:223–243

Sinani E, Meyer KE (2009) When and where does foreign direct investment generate positive spillovers. J Int Bus Stud 40(7):1075–1094

Singh J (2005) Collaborative networks as determinants of knowledge diffusion patterns. Manag Sci J 51(5):756–770, published by Institute for Operations Research and the Management Sciences

Sjöholm F (1996) International transfer of knowledge: the role of international trade and geographic proximity. Weltwirtschaftliches Archiv 132:97–115

Sjöholm F (1997) Knowledge inflow to Sweden: does geography and international trade matter? In: Fagerberg J, Hansson P, Lundberg L, Melchior A (eds) Technology and international trade. Edward Elgar, Cheltenham

Shulman S (1999) Owning the future. Houghton Mifflin, Boston

Slaughter S, Leslie LL (1997) Academic capitalism: politics, policies, and the entrepreneurial university. Johns Hopkins University Press, Baltimore

Slaughter S, Rhoades G (2004) Academic capitalism and the new economy: markets, state and higher education. The Johns Hopkins University Press, Baltimore

Smith PJ (1999) Do knowledge spillovers contribute to U.S. state output and growth? J Urban Econ 45:331–353

Smith V, Dilling-Hansen M, Eriksson T, Strøjer Madsen E (2000) R&D and productivity in Danish firms: some empirical evidence. CIE discussion papers 2000–05. Department of economics, Centre for industrial economics, University of Copenhagen

Smolny W (1998) Innovations, prices and employment. J Ind Econ 46:359–381

Soete L, Turner R (1984) Technology diffusion and the rate of technical change. Econ J 94 (385):612–623

Solow R (1950) Labor productivity functions in meat packing, The Review of Economics and Statistics, (8/1), MIT, Cambridge
Solow RM (1956) A contribution to the theory of economic growth. Q J Econ 70(1):65–94
Solvell O, Lindqvist G, Ketels Ch (2003) The cluster initiative greenbook. http://www.ictalent.org/_Code/UserFiles/Library/1.Cluster_Green_Book.pdf
Sonn JW, Storper M (2003)The increasing importance of geographical proximity in technological innovation: an analysis of U.S. patent citations, 1975–1997. Paper prepared for the conference: what do we know about innovation? Sussex, 13–15 November 2003 http://www2.lse.ac.uk/geographyAndEnvironment/whosWho/profiles/storper/pdf/Sonn_Storper.pdf
Sonn JW, Storper M (2005) The increasing importance of geographical proximity in technological innovation: an analysis of U.S. patent citations, 1975–1997. Mimeo
Stam E (2009) Cluster creation. Reg Stud 43(2):319–323
St K, Yegorov I (2000) Strategic alliances and technology transfer in central and eastern Europe. Sci Public Policy 27(4):265–274
Sternberg R (2000) Innovation networks and regional development – evidence from the European Regional Innovation Survey (ERIS). Eur Plann Stud J 8(4):389–407
Stilianos A, Tomkins J (2010) Technology adoption and club convergence. MPRA Paper 21137, University Library of Munich
Stokey NL (1988) Learning by doing and the introduction of new goods. J Polit Econ 96:701–717
Subramanian R, Scheff RP Jr, Quillinan JD, Wiper DS, Marsten RE (1994) Coldstart: fleet assignment at delta air lines. Interfaces 24:104–120
Stokke H (2008) Productivity growth and organizational learning. J Rev Dev Econ 12(4):764–778
Storper M (1997) The regional world: territorial development in a global economy. Guilford Press, New York
Storper M, Salais R (1997) Worlds of production: the action frameworks of the economy. Harvard University Press, Cambridge, MA
Storper M, Scott AJ (1987) High technology industry and regional development: a theoretical critique and reconstruction. Int Soc Sci J 112:215–232
Storper M, Venables AJ (2004) Buzz: face-to-face contact and the urban economy. J Econ Geogr 4 (4):351–370
Storper M, Walker R (1989) The capitalist imperative: territory, technology, and industrial growth. B. Blackwell, Oxford
Strambach S (1998) Knowledge-intensive business services (KIBS) as an element of learning regions – the case of Baden Württenberg. Paper presented to the 38th Congress of the European Regional Science Association, Vienna
Strebel J, Erdem T, Swait J (2004) Consumer search in high technology markets: exploring the use of traditional information channels. J Consum Psychol 14(1 & 2):96–104
Striukova L, Rayana T (2009) Public venture capital: missing link or weakest link? Int J Entrep Innov Manag 9(4):453–465
Stuart TE, Hoang H, Hybels RC (1999) Interorganizational endorsements and the performance of entrepreneurial ventures. Adm Sci Q 44(2):315–349
Surinach J, Moreno R, Vaya E (2007) Knowledge externalities, innovation clusters and regional development. Edward Elgar, Northampton
Sveikauskas L (1981) Technology inputs and multifactor productivity growth. Rev Econ Stat 63 (2):275–282
Swann GMP, Prevezer M, Stout DK (1998) The dynamics of industrial clustering: international comparisons in computing and biotechnology. Oxford University Press, Oxford/New York
Swann GMP (2005) High technology clusters: specialisation and interaction. In: Curzio QA, Fortis M (eds) Research and technological innovation: the challenge for a New Europe. Springer, Berlin, pp 129–150
Szlachta J, Zaucha J (2012) For an enhanced territorial dimension of the cohesion policy in Poland in the 2014–2020 period. Institute for Development, working papers (3). http://econpapers.repec.org/paper/irowpaper/1203.htm

Targetti F, Foti A (1997) Growth and productivity: a model of cumulative growth and catching-up. Camb, J Econ 21:27–43

Temple J (2002) An assessment of the new economy. CEPR Discussion Papers 3597, C.E.P.R. Discussion Papers

Thornhill S (2006) Knowledge, innovation and firm performance in high- and low-technology regimes. J Bus Ventur 21(5):687–703

Thumm N (2004) Strategic patenting in biotechnology. J Technol Anal Strateg Manag 16 (4):529–538

Tiits M, Kattel R, Kalvet T, Kaarli R (2003) Competitiveness of Estonian economy and future outlooks. Research and Development Council of Estonia, Tallinn

Tödtling F, Trippl M (2005) One size fits all? Towards a differentiated regional innovation policy approach. Res Policy 34:1023–1209

Tödtling F, Trippl M (2009) Innovation and knowledge links in metropolitan regions. The case of Vienna, SRE – discussion papers, 2009/02. Vienna University of Economics and Business, Vienna

Tondl G (2001) Convergence after divergence? Regional growth in Europe. Springer, Wien

Tondl G, Vuksic G (2003) What makes regions in Eastern Europe catching up? The role of foreign direct investment. Human resources and geography, Working paper 51. IEF, Vienna

Townsend J et al (1981) Science and technology indicators for the UK – innovations in Britain since 1945. SPRU occasional paper no. 16. University of Sussex Library

Traistaru I, Nijkamp IP, Resmini L (2003) The emerging economic geography in EU accession countries: concluding remarks and policy implications. In: Traistaru I, Nijkamp IP, Resmini L (eds) The emerging economic geography in EU accession countries. Ashgate, Aldershot, pp 442–449

Trajtenberg M (1990) A penny for your quotes: patent citations and the value of innovations. J Econ 21(1):172–187

Tripsas M, Gavetti G (2000) Capabilities, cognition and intertia: evidence from digital imaging. Harvard University, Boston

Turlea G, Bogdanowicz M (2007) EU enlargement: economic development and the information society. VUB Press, Brussels

Tushman ML, Anderson P (1986) Technological discontinuities and organizational environments. Adm Sci Q 31(3):439–465

Tushman ML, Romanelli E (1985) Organization evolution: a metamorphosis model of convergence and reorientation. In: Staw B, Cummings LL (eds) Research in organizational behavior, vol 7. JAI, Greenwich, pp 171–222

Umiński S (2002) Znaczenie zagranicznych inwestycji bezpośrednich dla transferu technologii do Polski. Published by Uniwersytet Gdański, Gdańsk

UNIDO (2004) Inserting local industries into global value chains and global production networks: opportunities and challenges for upgrading with a focus on Asia. Working Papers, Vienna. www.unido.org/resources/publications/. . .by-type/working-papers.html

Utterback JM, Abernathy WJ (1975) A dynamic model of process and product innovation. Int J Manag Sci 3(6):639–656

van Ark B, Piatkowski M (2004) Productivity, innovation and ICT, in old and New Europe. Research Memorandum GD-69. Groningen Growth and Development Centre, Groningen

van Ark B, Piatkowski M (2004a) Productivity, innovation and ICT in old and new Europe. Int Econ Econ Policy 1(2):215–246, Springer

van Ark B, Piatkowski M (2004b) Productivity, innovation and ICT in old and new Europe. Int Econ Econ Policy 1(2):215–246, Springer

van der Zwaan B et al (2002) Endogenous technological change in climate change modelling. International Institute for Applied Systems Analysis, Laxenburg

Vandenbussche J, Aghion P, Meghir C (2006) Growth, distance to frontier and composition of human capital. J Econ Growth 11:97–127

van Dijk SJ (2008) The illegitimacy of radical innovation in established high-tech companies: micro institutional conditions and strategic responses. Technische Universiteit Eindhoven, Eindhoven

Van Egeraat C, Curran D (2010) Defining and valuing the creative industries. Dublin City Council, Dublin

Van Everdingen Y, Waarts E (2003) The effect of national culture on the adoption of innovations. J Mark Lett 14(3):217–232, Springer

Van Looy B, Callaert J, Debackere K (2006) Scientific and entreprenurial activities within academia: mutually reinforcing, hampering or just co-existing? J Res Policy 34(3):321–342, Elsevier

van Meijl H (1995) Endogenous technological change: the case of information technology, Ph.D. dissertation, University of Limburg, Universitaire Press, Maastricht

Van Oort FG (2002) Agglomeration, economic growth and innovation: spatial analysis of growth and R&D externalities in the Netherlands. Erasmus University, Rotterdam

van Reenen J (1997) Employment and technological innovation: evidence from U.K. manufacturing firms. J Labor Econ 15:255–285

Vanhoudt P, Matha T, Smid B (2000) How productive are capital investments in Europe? EIB PAPERS, vol 5(2). European Investment Bank

Verspagen B (1993) Uneven growth between interdependent economies: an evolutionary view on technology gaps, trade and growth. Brookfield, Aldershot

Verspagen B (1997) Measuring intersectoral technology spillovers: estimates from the European and US patent office databases. Econ Syst Res 9:47–65

Verspagen B (2008) Knowledge flows, patent citations and the impact of science on technology. Econ Syst Res J 20:339–366, Taylor & Francis

Viale R, Orto S (1998) Neocorporatist or evolutionary triple helix? Suggesions coming from European regions, Sci Public Policy J 63, http://www.sociologiadip.unimib.it/dipartimento/ricerca/pdfDownload.php?idPaper=63

Vivarelli M, Pianta M (2000) The employment impact of innovation: evidence and policy. Routledge, London, p 240

Vogel J (2012) The two faces of R&D and human capital: evidence from Western European regions. Economics series working papers 599. Department of Economics, University of Oxford

von Hippel E (1986) Cooperation between competing firms: informal know-how trading. MIT Press, Cambridge, MA

von Hippel E (1998) Economics of product development by users: the impact of 'sticky' local information. Manag Sci 44(5):629–644

Vonortas NS (2003) Technology licensing, organization for economic cooperation and development. OECD, Paris, October

Walker H, Harland C, Knight L, Lamming R (2005) Outsourcing: assessing the risks and benefits for organizations, sectors and nations. Int J Oper Prod Manag 25(9):831–850

Wallsten SJ (2000) The R&D boondoggle: why is the government subsidizing commercially promising business projects? Regul J 23(4):12–16

Walsh JP, Arora A, Cohen WM (2003) The patenting and licensing of research tools and biomedical innovation. In: Cohen WM, Merrill S (eds) Patents in the knowledge-based economy. National Academies Press, Washington, DC, pp 285–340

Walsh JP, Cohenb WM, Cho C (2007) Where excludability matters: material versus intellectual property in academic biomedical research. J Res Policy 36:1184–1203, Elsevier

Watts RJ, Porter AL (1997) Innovation forecasting. Technol Forecast Soc Change 56:25–47

Webster A (ed) (1996) Building new bases for innovation: the transformation of the R&D system in post-socialist states. Anglya Politechnic University, Cambdrige

Weresa MA (2002) Wpływ handlu zagranicznego i inwestycji bezpośrednich na innowacyjność polskiej gospodarki. Published by SGH Warsaw

Weresa M (2007) Formy i metody powiązań nauki i biznesu. In: Weresa M (ed) Transfer wiedzy z nauki do biznesu. Doświadczenia regionu Mazowsze, Oficyna Wyd. SGH, Warszawa

Weresa MA, Leal Filho W (2008) Fostering innovation and knowledge transfer in European regions. Peter Lang, Bruxelles

Wevers RC, van Beers C, Zand F (2008) R&D cooperation diversity and innovation. Paper presented at DRUID conference 2008, June 17–20, Copenhagen. www2.druid.dk/conferences/viewpaper

Wilson DJ (2001) Is embodied technology the result of upstream R&D? Industry-level evidence. Working papers in applied economic theory, 2001–17, Federal Reserve Bank of San Francisco
Wintjes R, Hollanders H (2011) Innovation pathways and policy challenges at the regional level: smart specialization, UNU-MERIT Working Papers. UNU-MERIT, Maastricht
Witt U, Zellner C (2007) How firm organizations adapt to secure a sustained knowledge transfer. Max Planck Institute of Economics, Jena
Wolfe DA (2000) Social capital and cluster development in leraning regions. In: Holbrook A, Wolfe DA (eds) Knowledge, clusters and learning regions. School of Policy Studies, Queens University, Kingston
Wolfe DA, Bramwell A (2008) Innovation, creativity and governance: social dynamics of economic performance in city regions. Policy Pract 10(2–3):170–182
Wolff EN, Nadiri MI (1993) Spillover effects, linkage structure, and research and development. Struct Change Econ Dyn 4:315–331
Worgan A, Nunn S (2002) Exploring a complicated labyrinth: some tips on using patent data to measure urban and regional innovation. Econ Dev Q 16:229–236
Wu W (2005) Dynamic cities and creative clusters, World Bank Policy Research Working Paper 3509
Wunderlich S, Khalil T (2002) Introducing technology timeline interpretation to technology difussion and maturity analysis as applied to different industrial sectors. University of Miami, Miami
Yu D (1991) Achieving excellence in the global marketplace using knowledge-based simulation. In: Proceedings of the first international conference on AI applications on wall street. IEEE Computer Society Press, Washington, DC, pp 103–108
Zahra SA, George G (2002) International entrepreneurship: the current status of the field and future research agenda. In: Hitt MA (ed) Strategic entrepreneurship: creating a new mindset. Blackwell, Malden, pp 255–288
Zaucha J (2007) Rola przestrzeni w kształtowaniu relacji gospodarczych. Ekonomiczne fundamenty planowania przestrzennego w Europie Bałtyckiej. University of Gdansk, Gdańsk
Zemlickiene V (2011) Analysis of high-technology product development models. Intellect Econ 5 (2 (10)):283–297
Zhou KZ, Yim CK, Tse DK (2005) The effects of strategic orientations on technology- and market-based breakthrough innovations. J Mark 69(2):42–60, Emerald Pub
Zook MA (2005) The geography of the internet industry. Blackwell, Oxford
Zucker LG, Darby MR, Armstrong J (1994) Working Paper No. 4946, National Bureau of Economic Research, Cambridge
Zucker LG, Darby MR, Brewer MB (1994b) Intellectual capital and the birth of U.S. biotechnology enterprises, vol 4653, National Bureau of Economic Research Working Paper. National Bureau of Economic Research, Cambridge
Zucker LG, Darby MR (1995) Present at the revolution: transformation of technical identity for a large incumbent pharmaceutical firm after the biotechnological breakthrough, National Bureau of Economic Research Working Paper No. 5243, August
Zucker LG, Darby MD, Armstrong J (1998) Geographically localized knowledge: spillovers or markets? Econ Inq 36(1):65–86

Web Sources

Abernathy WJ, Utterback JM. Innovation over time and in historical context: pattern of industrial innovation. www.ec.unipg.it/. . ./6_%20Abernathy%20-%20Utterback%20-Innovation%20over%20time%20and%20in%20historic
Asheim B, Boschma R, Cooke P (2007) Constructing regional advantage: platform policies based on related variety and differentiated knowledge bases. Working paper. CIRCLE, Lund. http://econ.geo.uu.nl/boschma/artashboscooke2.pdf

Asheim B, Coenen L, Moodysson J, Vang J. Regional innovation system policy: a knowledge-based approach. Lund working paper 2005/13. http://www.circle.lu.se/upload/CIRCLE/workingpapers/200513_Asheim_et_al.pdf
Autant-Bernard C (2006) R&D lab location. Evidence from the French case, CREUSET – University of Saint-Etienne. http://hal.archives-ouvertes.fr/docs/00/06/65/23/PDF/Autant_CESPRI.pdf
ftp://zappa.ubvu.vu.nl/20080013.pdf
http://en.wikipedia.org/wiki/Health_hazards_in_semiconductor_manufacturing_occupations
http://failuremag.com/index.php/feature/article/eye_on_lasik/P2/#ixzz1Y1M98FmT
http://www.microvision.com/pdfs/benefits_apps.pdf
http://www.commerce.wa.gov.au/worksafe/content/safety_topics/Plant_and_machines/Additional_information/Laser_hazards.html#3.%20Hazard%20identification
http://www.feweb.vu.nl/ersa2005/final_papers/381.pdf
http://www.uni-marburg.de/fb19/personal/professoren/brenner/wp10.pdf
http://www.ikinet.uniroma2.it/docs2811/REPORT_I.pdf
http://www.tandfonline.com/doi/pdf/10.1080/09654310802682115
http://www2.druid.dk/conferences/viewpaper.php?id=2701&cf=18
http://www.circle.lu.se/upload/CIRCLE/workingpapers/200815_Moodysson_et_al.pdf
http://www.utoronto.ca/isrn/publications/WorkingPapers/Working02/Mytelka02_Clustering.pdf
http://www2.druid.dk/conferences/viewpaper.php?id=278&cf=8
http://eprints.nuim.ie/1906/1/WP_Van_Egeraat_Curran_Biotech%2520Network_NIRSA.pdf
http://oaithesis.eur.nl/ir/repub/asset/8899/MA-thesis%20Jelte-Dijkstra.pdf
http://www2.druid.dk/conferences/viewpaper.php?id=2281&cf=16
http://www.schumpeter2010.dk/index.php/schumpeter/schumpeter2010/paper
http://epub.wu.ac.at/550/1/document.pdf
Kearney AT, Inc (2005) Outsourcing strategically for sustainable competitive advantage. Centre for Strategic Supply Research. http://www.capsresearch.org/publications/pdfs-protected/monczka2005.pdf
OECD (2009) The impact of the crisis on ICT and ICT-related employment. http://www.oecd.org/internet/interneteconomy/43969700.pdf
Plunket A, Intra-sector and cross-region analysis of clusters, agglomeration and spillovers in high technology sectors. http://www.oecd.org/dataoecd/22/17/37443555.pdf
Ponds R, Oort F, Frenken K (2009) Innovation, spillovers, and university-industry collaboration: an extended knowledge production function approach. http://ideas.repec.org/p/egu/wpaper/0903.html. Accessed Feb 2009
Risks and Benefits of Computer Technology, eHow.com. http://www.ehow.com
Shapira P (1996) An overview of technology diffusion policies and programs to enhance the technological absorptive capabilities of small and medium enterprises. School of public policy, Georgia Institute of Technology, Atlanta. http://www.prism.gatech.edu/~jy5/pubs/oecdtech.htm. Accesed Aug 1996
The Advantages of Semiconductors, eHow.com. http://www.ehow.com
The Disadvantages of Technology on Communication, eHow.com. http://www.ehow.com
Thulin P (2009) Labor mobility, knowledge diffusion and regional growth. CESIS – Centre of excellence for science and innovation studies, Royal institute of technology, CESIS electronic working paper series, paper no. 209. http://cesis.abe.kth.se/documents/CESISWP209.pdf. Accessed Dec 2009
What Are the Benefits of Communication Technology? eHow.com. http://www.ehow.com

List of Interviewees

Prof. Patrick Hendrick (Université libre de Bruxelles), Joseph Bernier (Liege Science Park), Prof. József Rohács (Department of Aircraft and Ships, Budapest University of Technology), Dr. Daniel Hanus (Department of Automotive and Aerospace Engineering, Czech Technical University in Prague), Prof. Romana Śliwa (Rzeszow University of Technology, Centre of Advanced Technologies), Prof. Vladimir Marik (Department of Cybernet, Czech Technical University), Prof. Jüri Vain (Department of Computer Science, Talin TechnicalUniversity), Dr. János Levendovszky (Faculty of Electrical Engineering and Informatics, Budapest University of Technology and Economics), Dr. Rudy Dekeyser (the Flanders Institute for Biotechnology), Dr. Laurent Corbesier (Forem Formation Biotechnologie), Prof. Indrikis Muiznieks (Department of Microbiology and Biotechnology, University of Latvia), Prof. Károly Marialigeti (Institute of Biology, Eötvös Lóránd University), Prof. George Szekeres (Department of Biotechnology Innovation, Pécs Industrial Park), Prof. Wanda Dobryszycka (Faculty of Pharmacy, Wroclaw Medical University), Prof. Grażyna Lewandowicz (Poznan University of Life Science), Marijke Lein (Flanders Institute for Biotechnology), Pascale Philips (BioScience, Zwijnaarde Science Park), Prof. Krzysztof Staroń (Faculty of Molecular Biology of Warsaw University), Prof. Eerik Lossmann (Faculty of Information Technology, Tallin University of Technology), Prof. Szczepanski Stanislaw (Faculty of Electronics, Telecommunications and Informatics, Gdansk Technical University), Prof. Enn Mellikov (Department of Materials Science, Tallinn University of Technology), Frederik Cambier (Departement Micro-Technologies, Liège Science Park), Prof. Didier Georges (Grenoble Institute of Technology), Prof Zbigniew Lisik (Department of Semiconductor and Optoelectronics Devices, Lodz Technology University), Prof. Krzysztof Kubiak (Faculty of Mechanical Engineering and Aeronautics, Rzeszow University of Technology), Prof. Rimantas Kanapenas (Vilnius Laser Technology Center), Dr. Marc Georges (Laser Techniques Activity, Université de Liège), Prof. Michele Cincera (Université libre de Bruxelles), Prof. Dr. Ferenc Krausz (Max-Planck-Institut für Quantenoptik), Prof. Riku Jäntti (Department of Communications and Networking, Aalto University), Prof. Peter Göhner (Institute of Industrial Automation and Software Engineering, Universität Stuttgart).

M. Runiewicz-Wardyn, *Knowledge Flows, Technological Change and Regional Growth in the European Union*, Contributions to Economics,
DOI 10.1007/978-3-319-00342-9, © Springer International Publishing Switzerland 2013

Annex 1

Table 1 Manufacturing industries classified according to their global technological intensity (ISIC revision 2 and NACE revision 1.1) OECD classification

High-technology 1.1	NACE revision	ISIC revision 2
1. Aerospace	35.3	3845
2. Computers, office machinery	30	3825
3. Electronics-communications	32	3832
4. Pharmaceuticals	24.4	3522
5. Scientific instruments	33	385
Medium-high technology		
6. Motor vehicles	34	3843
7. Electrical machinery	31	383–3832
8. Chemicals	24–24.4	351 + 352–3522
9. Other transport equipment	35.2 + 35.4 + 35.5	3842 + 3844 + 3849
10. Non-electrical machinery	29	382–3825
Medium-low technology		
11. Rubber and plastic products	25	355 + 356
12. Shipbuilding	35.1	3841
13. Other manufacturing	36.2 through 36.6 39	39
14. Non-ferrous metals	27.4 + 27.53/54	372
15. Non-metallic mineral products	26	36
16. Fabricated metal products	28	381
17. Petroleum refining	23	351 + 354
18. Ferrous metals	27.1 through 27.3 + 27.51/52	371
Low-technology		
19. Paper printing	21 + 22	34
20. Textile and clothing	17 through 19	32
21. Food, beverages, and tobacco	15 + 16	31
22. Wood and furniture	20 + 36.1	33

Source: High-technology and knowledge-intensive sectors, December 2011, Eurostat

M. Runiewicz-Wardyn, *Knowledge Flows, Technological Change and Regional Growth in the European Union*, Contributions to Economics,
DOI 10.1007/978-3-319-00342-9, © Springer International Publishing Switzerland 2013

Table 2 Knowledge-intensive high-technology service and knowledge-intensive service industries (ISIC revision 2 and NACE revision Eurostat classification)

Knowledge-intensive services	NACE revision	ISIC revision 3
Water transport	61	611 + 612
Air transport	62	621 + 622
Post and telecommunications	64	641 + 642
Financial intermediation, except insurance and pension funding	65	651 + 659
Knowledge-intensive high-technology services		
Post and telecommunications	64	641 + 642
Computer and related activities	72	721 + 722 + 723 + 724 + 725 + 729
Research and development	73	731 + 732
Insurance and pension funding, except compulsory social security	66	660
Activities auxiliary to financial intermediation	67	671 + 672
Real estate activities	70	701 + 702
Renting of machinery and equipment without operator and of personal and household goods	71	711 + 712 + 713
Computer and related activities	72	721 + 722 + 723 + 724 + 725 + 729
Research and development	73	731 + 732
Other business activities	74	741 + 742 + 743 + 749
Education	80	801 + 802 + 803 + 809
Health and social work	85	851 + 852 + 853
Recreational, cultural, and sporting activities	92	921 + 922 + 923 + 924

Source: High-technology and knowledge-intensive sectors, December 2011, Eurostat

Annex 2

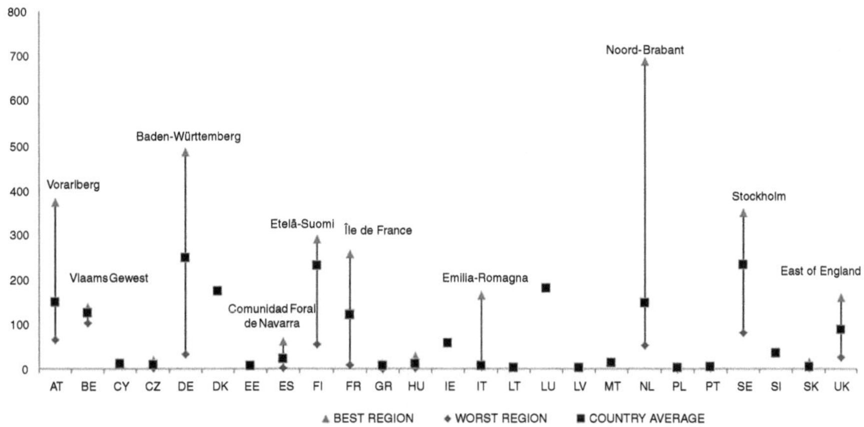

Fig. 1 Patent applications to the EPO per 1 mln (average 1996/97/98-2007) (Note: Minimum values are indicated by a *quadrangle*, maximum values by a *star* and the national average by a *dash*. Source: own calculations based on data from Eurostat – REGIO)

M. Runiewicz-Wardyn, *Knowledge Flows, Technological Change and Regional Growth in the European Union*, Contributions to Economics,
DOI 10.1007/978-3-319-00342-9, © Springer International Publishing Switzerland 2013

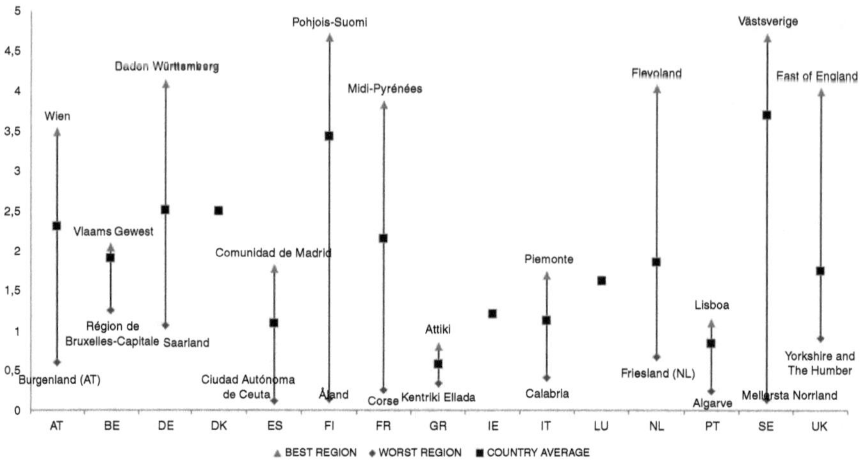

Fig. 2 EU-15 total R&D as a percentage of GDP (average 2001–2007) (Note: Minimum values are indicated by a *quadrangle*, maximum values by a *star*, and the national average by a *dash*. Source: own calculations based on data from Eurostat – REGIO)

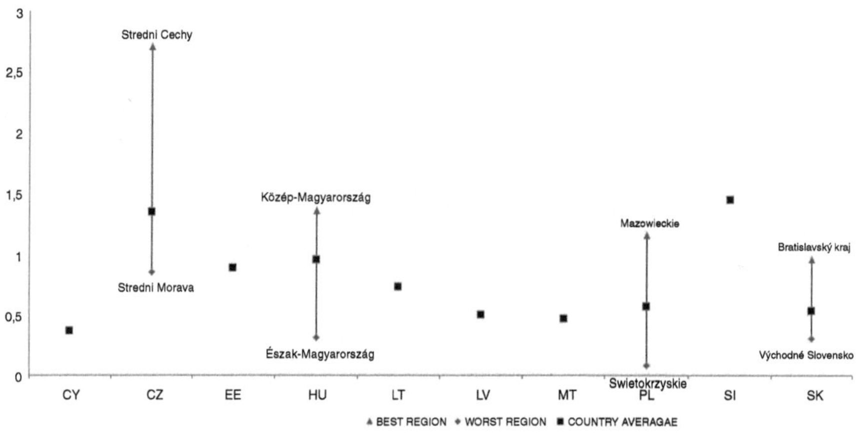

Fig. 3 EU-10 total R&D as a percentage of GDP (average 2001–2007) (Note: Minimum values are indicated by a *quadrangle*, maximum values by a *star*, and the national average by a *dash*. Source: own calculations based on data from Eurostat – REGIO)

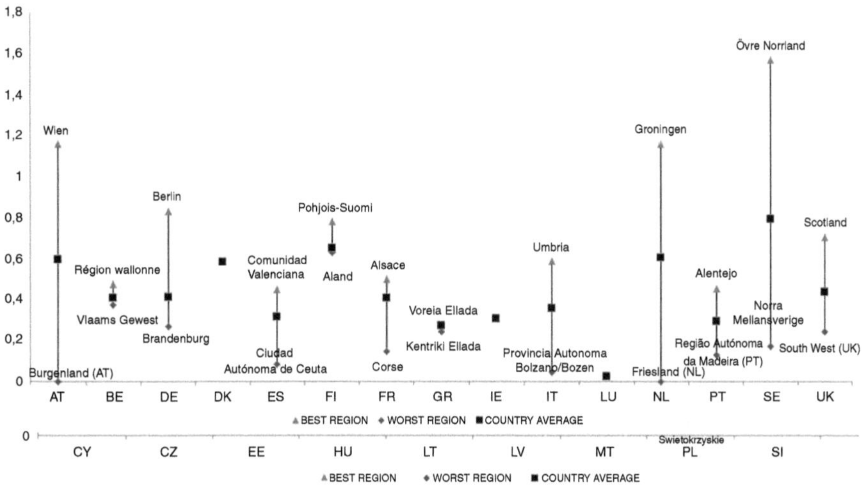

Fig. 4 EU-15 university R&D as a percentage of GDP (average 2001–2007) (Note: Minimum values are indicated by a *quadrangle*, maximum values by a *star*, and the national average by a *dash*. Source: own calculations based on data from Eurostat – REGIO)

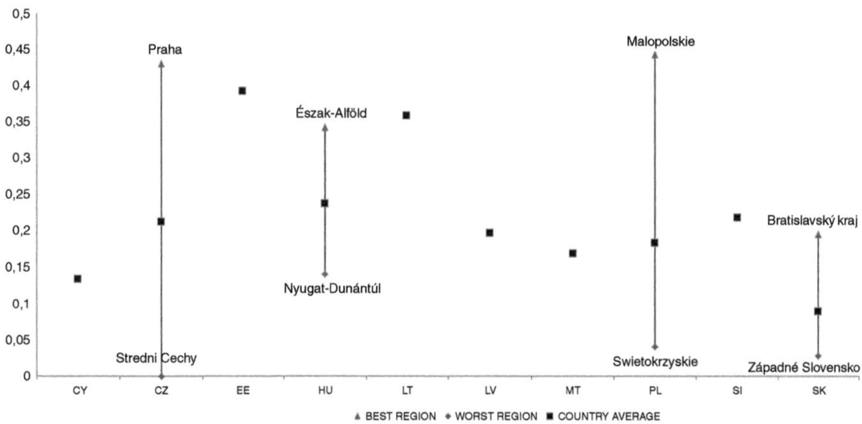

Fig. 5 EU-10 university R&D as a percentage of GDP (average 2001–2007) (Note: Minimum values are indicated by a *quadrangle*, maximum values by a *star*, and the national average by a dash. Source: own calculations based on data from Eurostat – REGIO)

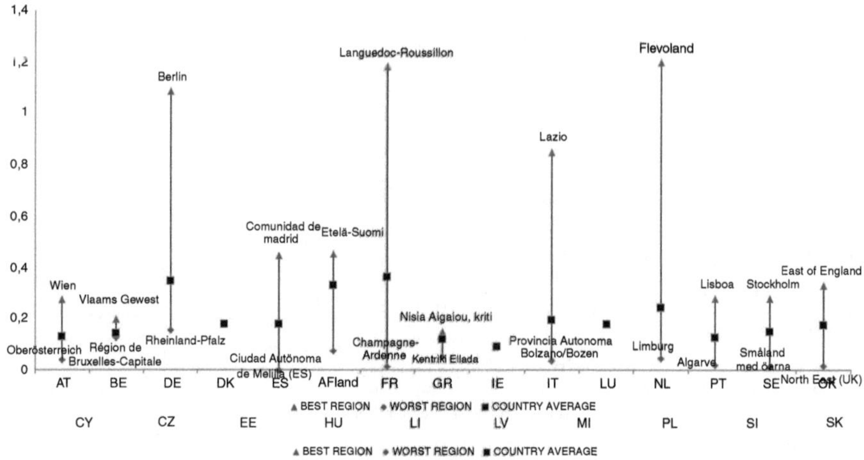

Fig. 6 EU-15 government R&D as a percentage of GDP (average 2001–2007) (Note: Minimum values are indicated by a *quadrangle*, maximum values by a *star*, and the national average by a *dash*. Source: own calculations based on data from Eurostat – REGIO)

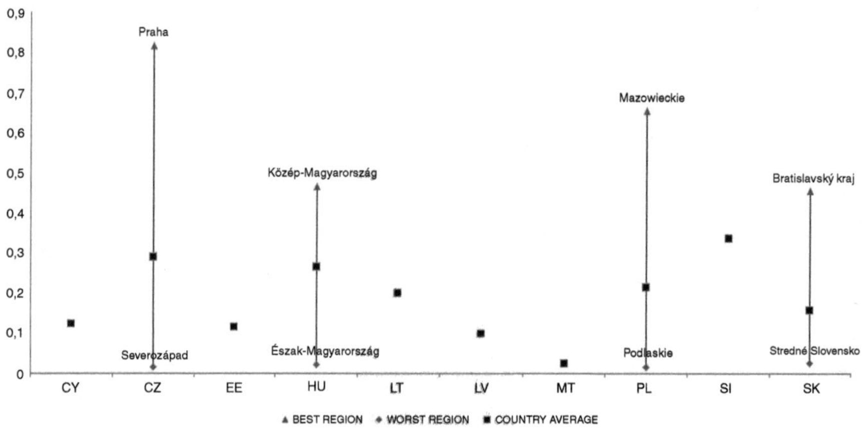

Fig. 7 EU-10 government R&D as a percentage of GDP (average 2001–2007) (Note: Minimum values are indicated by a *quadrangle*, maximum values by a *star*, and the national average by a *dash*. Source: own calculations based on data from Eurostat – REGIO)

Annex 3: The Survey Study Results

Scientific field		Aerospace	Biotechnology	Communication	Computer	Semiconductors	Laser technology
A. Number of respondents		5	10	5	5	5	5
(a) Major source of R&D funding (from 1 to 5, where 1 – very significant and 5 – least significant)	University	3(4)	5 (2)	1(2)	2(2)	2(3)	1(5)
		1(3)	4 (3)	1(3)	1(4)	2(4)	1(3)
		1(2)	1 (4)			1(2)	
	Industry/business	2(1)	4 (1)	2(3)	1(1)	4(1)	1(2)
		2(3)	3 (2)		2(3)	1(2)	1(4)
		1(2)	3 (4)				
	Government	3(2)	6 (1)	3(1)	1(1)	4(1)	1(1)
		1(1)	2 (2)		2(2)	1(2)	1(2)
		1(3)	2 (4)				
	EU funds	2(1)	4 (2)	2(2)	1(2)	2(3)	1(5)
		1(2)	3 (3)	1(5)	1(3)	1(2)	1(1)
		1(5)	3 (4)		1(4)	1(3)	
		1(4)				1(4)	
	Other (venture capital, etc.)	1(3)	6 (5)	1(4)	1(3)	2(3)	1(5)
			2 (1)		1(5)	2(4)	
B. Main channels of knowledge diffusion							
(a) What are your university/R&D lab's most common forms of knowledge diffusion? (from 1 to 5, where 1 – very significant and 5 – least significant)	Seminars, conferences, workshops	2(2)	1 (2)	2(2)	2(2)	2(1)	1(2)
		1(3)	4 (3)		1(3)	2(2)	
		1(4)	1 (4)			1(3)	
		1(1)	2 (5)				
	Collaboration with business sector	3(1)	3 (1)	2(1)	3(1)	3(1)	1(3)
		1(5)	2 (2)	1(3)		1(3)	
		1(2)	5 (3)				

(continued)

M. Runiewicz-Wardyn, *Knowledge Flows, Technological Change and Regional Growth in the European Union*, Contributions to Economics,
DOI 10.1007/978-3-319-00342-9, © Springer International Publishing Switzerland 2013

Scientific field		Aerospace	Biotechnology	Communication	Computer	Semiconductors	Laser technology
	Publishing activity	3(2)	2 (1)	1(3)	2(2)	2(2)	1(2)
		2(1)	4(4)			2(1)	
			2 (5)			1(4)	
	Patents	–	3 (3)		–		2(1)
			1 (1)				
	Informal (face-to-face) contact	2(4)	1(1)	3(3)	2(3)	4(2)	1(4)
		1(2)	4(2)				
		1(3)	1(3)				
		1(1)	3(4)				
C. Geographically and technologically mediated knowledge spillovers							
(a) The role of local (intra-regional) source knowledge for innovative activity	Significant	5	10	2	3	5	1
	Insignificant	0	0	1	0		1
(b) Significance of local (intra-regional) source of knowledge by type of knowledge	Technological (T)	5	8	3	3	5	2
	Market (MT)	1	5	0	2	1	0
	Managerial (ML)	2	3	0	2	0	0
	Other						
(c) The significance of the other sources of knowledge (inter-, extra- regional and global? By which type of knowledge (technological, market, managerial?	European (inter-regional)	3(T)	3 (T)	2(T)	1(MT)	2(T)	0
		3(MT)	2 (MT)	1(MT)			
		3 (ML)					
	European (extra regional)	5(T)	4 (T)	3(T)	1(T)	5(T)	2
					1(MT)	5(MT)	
	Global	5(T)	9 (T)	3(T)	2(T)	5(T)	2
						5(MT)	

D. Knowledge and technological networks							
The role of location for the knowledge networking? (from 1 to 5, where 1 – very significant and 5 – least significant)	Home institution and in the home region institutions	3(3) 1(2)	5(1) 3 (3)	3(1)	1(2) 2(3)	4(1) 1(4)	1
	Other institutions in my country	2(1) 1(1)	4 (3)	2(2)	2(2)	4(1) 1(2)	0
	Other institutions in EU	3(1)	3 (3)	1(2) 1(3)	3(1)	3(2) 1(3) 1(4)	2
	Other institutions in other countries	2(1)	–	–	2(1)	3(2) 1(3) 1(4)	2
E. Secrecy in university research							
(a) Have you requested any research results/r materials (software, genetic sequences, data) since 2009 from other scientists but did not get access?	Yes						
	No	5	10	3	3	5	2
(b) Have you denied other scientists access to your research results or materials since 2009?	Yes		2				
	No	5	8	3	3	5	2

Source: own elaboration

MIX
Papier aus verantwortungsvollen Quellen
Paper from responsible sources
FSC® C105338

If you have any concerns about our products,
you can contact us on
ProductSafety@springernature.com

In case Publisher is established outside the EU,
the EU authorized representative is:
Springer Nature Customer Service Center GmbH
Europaplatz 3, 69115 Heidelberg, Germany

Printed by Libri Plureos GmbH
in Hamburg, Germany